Springer Theses

Recognizing Outstanding Ph.D. Research

Aims and Scope

The series "Springer Theses" brings together a selection of the very best Ph.D. theses from around the world and across the physical sciences. Nominated and endorsed by two recognized specialists, each published volume has been selected for its scientific excellence and the high impact of its contents for the pertinent field of research. For greater accessibility to non-specialists, the published versions include an extended introduction, as well as a foreword by the student's supervisor explaining the special relevance of the work for the field. As a whole, the series will provide a valuable resource both for newcomers to the research fields described, and for other scientists seeking detailed background information on special questions. Finally, it provides an accredited documentation of the valuable contributions made by today's younger generation of scientists.

Theses may be nominated for publication in this series by heads of department at internationally leading universities or institutes and should fulfill all of the following criteria

- They must be written in good English.
- The topic should fall within the confines of Chemistry, Physics, Earth Sciences, Engineering and related interdisciplinary fields such as Materials, Nanoscience, Chemical Engineering, Complex Systems and Biophysics.
- The work reported in the thesis must represent a significant scientific advance.
- If the thesis includes previously published material, permission to reproduce this must be gained from the respective copyright holder (a maximum 30% of the thesis should be a verbatim reproduction from the author's previous publications).
- They must have been examined and passed during the 12 months prior to nomination.
- Each thesis should include a foreword by the supervisor outlining the significance of its content.
- The theses should have a clearly defined structure including an introduction accessible to new PhD students and scientists not expert in the relevant field.

Indexed by zbMATH.

More information about this series at https://link.springer.com/bookseries/8790

Masataka Mogi

Quantized Phenomena of Transport and Magneto-Optics in Magnetic Topological Insulator Heterostructures

Doctoral Thesis accepted by
The University of Tokyo, Tokyo, Japan

Author
Dr. Masataka Mogi
Department of Applied Physics
University of Tokyo
Tokyo, Japan

Supervisor
Prof. Yoshinori Tokura
Department of Applied Physics
University of Tokyo
Tokyo, Japan

RIKEN Center for Emergent Matter
Science
Saitama, Japan

ISSN 2190-5053   ISSN 2190-5061   (electronic)
Springer Theses
ISBN 978-981-19-2139-1   ISBN 978-981-19-2137-7   (eBook)
https://doi.org/10.1007/978-981-19-2137-7

This Springer imprint is published by the registered company Springer Nature Singapore Pte Ltd.
The registered company address is: 152 Beach Road, #21-01/04 Gateway East, Singapore 189721,
Singapore

Supervisor's Foreword

Since the discovery of the quantum Hall effect, the importance of topological aspects in material properties has been increasing in contemporary condensed-matter physics. The concept of topological materials has allowed us to recognize the emergence of edge and surface states, whose conduction is essentially non-dissipative and robust against disorders. Quantization phenomena of physical responses are the hallmarks of such topological states and the exploration of them has attracted much attention in terms of both fundamental research and technological applications. To this end, time-reversal symmetry broken magnetic Topological Insulator (TI), which features massive Dirac fermions on the surface states while the interior is gapped and insulating, provides an ideal platform to explore unique quantization phenomena. The quantum anomalous Hall effect is one striking phenomenon, which generates non-dissipative chiral edge conduction without external magnetic fields.

Dr. Masataka Mogi's thesis presents electrical transport and magneto-optical studies in magnetic TIs by engineering their heterostructures. He made many breakthroughs in quantized phenomena relevant to the quantum anomalous Hall effect, such as enhancing the observable temperature and discoveries of new phenomena. To achieve them, he has developed a film-growth technique to control the magnitude and direction of magnetic moments and the spatial degrees of freedom in the TI surfaces. He has succeeded in making the surface magnetizations to fully point outwards/inwards, resulting in the emergence of an exotic insulating state, now known as the axion insulator state. Furthermore, he has developed the corroborated studies of state-of-the-art experimental techniques of precise electrical transport and low-energy (terahertz) magneto-optical spectroscopy. He demonstrated the half-quantized Hall effect originating from a single species of Dirac fermions by measuring the half-quantized Faraday and Kerr rotations and by directly comparing with Hall conductivity. This is a key result for establishing the parity anomaly in condensed-matter systems, which is the term originally developed in quantum field theory on two-dimensional space. This thesis provides the fundamentals of unconventional electromagnetic responses in magnetic TIs. These results may pave the way for designing novel device applications based on non-dissipative topological electronic properties.

Tokyo, Japan   Yoshinori Tokura
May 2020

Preface

In crystalline solids, band theory successfully classifies metals, semiconductors, and insulators, and reveals the basic electrical and optical properties via the characterization of electronic states described by energy-momentum relations, namely the electronic band structures. Recent significant progress in band theory is to capture the electronic states in terms of a mathematical concept of topology, clarifying that materials can possess novel edge or surface electronic modes that do not appear in their bulk band structures. This new phase of matter is called a topological phase, which is distinct from ordered phases arising from spontaneous symmetry breaking in that a non-trivial topological number is defined by 'global' electronic structures. Such topological phases are robust against the details of materials properties inevitably containing 'local' disorders such as defects and/or impurities.

The topological edge and surface modes can give rise to quantization phenomena in transport and optical properties, such as the quantized Hall resistance in the quantum Hall and quantum spin Hall effects. Because these quantization phenomena can accompany transport properties without energy dissipation and/or unusual quantum statistics owing to the emergence of charge-neutral Majorana fermions, intensive theoretical and experimental efforts have been devoted to discovering new quantization phenomena. A three-dimensional (3D) Topological Insulator (TI) is one of the typical topological phases guaranteed by time-reversal symmetry, which is electrically insulating in its interior while the surface is conducting owing to the topologically non-trivial band structures. Interestingly, the metallic surface state is characterized by spin-polarized two-dimensional massless Dirac fermions. By exploiting the relativistic quantum nature of the Dirac fermions, various novel quantum phenomena can be expected to appear.

In this thesis, we present the exploration of various quantized phenomena of transport and magneto-optics in 3D TIs. By using molecular-beam epitaxy film-growth techniques, we designed heterostructures based on 3D TIs endowed with magnetism. We first study the quantum anomalous Hall effects in TI films with magnetic modulation doping (Chap. 3), which enhances the stability, and also with magnetic proximity effect (Chap. 4). These studies enable us to further explore topological phase transitions to a trivial insulator phase and also a new topological phase of an axion insulator (Chap. 5). Furthermore, we discover a new fractional (half-integer) quantum Hall effect on the surface of a 3D TI, which demonstrates the parity anomaly in terms of quantum field theory (Chap. 6). We believe that these results would be a basis for future technologies such as ultra-low-energy consumption electronic devices and fault-tolerant topological quantum computers.

Tokyo, Japan   Masataka Mogi
December 2021

Parts of this thesis have been published in the following journal articles:

1. **Magnetic modulation doping in topological insulators toward higher-temperature quantum anomalous Hall effect**
   M. Mogi, R. Yoshimi, A. Tsukazaki, K. Yasuda, Y. Kozuka, K. S. Takahashi, M. Kawasaki & Y. Tokura
   Applied Physics Letters **107**, 182401 (2015).

2. **Terahertz spectroscopy on Faraday and Kerr rotations in a quantum anomalous Hall state**
   K. N. Okada, Y. Takahashi, M. Mogi, R. Yoshimi, A. Tsukazaki, K. S. Takahashi, N. Ogawa, M. Kawasaki & Y. Tokura
   Nature Communications **7**, 12245 (2016).

3. **A magnetic heterostructure of topological insulators as a candidate for an axion insulator**
   M. Mogi, M. Kawamura, R. Yoshimi, A. Tsukazaki, Y. Kozuka, N. Shirakawa, K. S. Takahashi, M. Kawasaki & Y. Tokura
   Nature Materials **16**, 516 (2017).

4. **Tailoring tricolor structure of magnetic topological insulator for robust axion insulator**
   M. Mogi, M. Kawamura, A. Tsukazaki, R. Yoshimi, K. S. Takahashi, M. Kawasaki & Y. Tokura
   Science Advances **3**, eaao1669 (2017).

5. **Quantized chiral edge conduction on domain walls of a magnetic topological insulator**
   K. Yasuda, M. Mogi, R. Yoshimi, A. Tsukazaki, K. S. Takahashi, M. Kawasaki, F. Kagawa & Y. Tokura
   Science **358**, 1311 (2017).

6. **Ferromagnetic insulator $Cr_2Ge_2Te_6$ thin films with perpendicular remanence**
   M. Mogi, A. Tsukazaki, Y. Kaneko, R. Yoshimi, K. S. Takahashi, M. Kawasaki & Y. Tokura
   APL Materials **6**, 091104 (2018).

7. **Topological quantum phase transition in magnetic topological insulator upon magnetization rotation**
   M. Kawamura, M. Mogi, R. Yoshimi, A. Tsukazaki, Y. Kozuka, K. S. Takahashi, M. Kawasaki & Y. Tokura
   Physical Review B **98**, 140404(R) (2018).

8. **Large anomalous Hall effect in topological insulators with proximitized ferromagnetic insulators**
M. Mogi, T. Nakajima, V. Ukleev, A. Tsukazaki, R. Yoshimi, M. Kawamura, K. S. Takahashi, T. Hanashima, K. Kakurai, T. Arima, M. Kawasaki & Y. Tokura
Physical Review Letters **123**, 016804 (2019).

9. **Quantum anomalous Hall effect driven by magnetic proximity coupling in all-telluride based heterostructure**
R. Watanabe, R. Yoshimi, M. Kawamura, M. Mogi, A. Tsukazaki, X. Z. Yu, K. Nakajima, K. S. Takahashi, M. Kawasaki & Y. Tokura
Applied Physics Letters **115**, 102403 (2019).

10. **Current-induced switching of proximity-induced ferromagnetic surface states in a topological insulator**
M. Mogi, K. Yasuda, R. Fujimura, R. Yoshimi, N. Ogawa, A. Tsukazaki, M. Kawamura, K. S. Takahashi, M. Kawasaki & Y. Tokura
Nature Communications **12**, 1404 (2021).

11. **Experimental signature of parity anomaly in semi-magnetic topological insulator**
M. Mogi, Y. Okamura, M. Kawamura, R. Yoshimi, K. Yasuda, A. Tsukazaki, K. S. Takahashi, T. Morimoto, N. Nagaosa, M. Kawasaki, Y. Takahashi & Y. Tokura
Nature Physics **18**, 390 (2022).

Acknowledgements

I sincerely would like to give my gratitude to Prof. Yoshinori Tokura for his continuous guidance and encouragement throughout my graduate studies.

I would like to appreciate the following people for the many big collaborations: Profs. Masashi Kawasaki, Naoto Nagaosa, Taka-hisa Arima, Youtarou Takahashi, Takahiro Morimito, Dr. Kenji Yasuda (now in MIT), Ms. Reika Fujimura, Mr. Ryota Watanabe, Dr. Yoshihiro Okamura, Dr. Ken Okada, and Prof. Yusuke Kozuka (now in NIMS) in University of Tokyo; Dr. Minoru Kawamura, Dr. Ryutaro Yoshimi, Dr. Kei Takahashi, Dr. Yoshio Kaneko, Prof. Taro Nakajima (now in UTokyo), Dr. Victor Ukleev (now in PSI), Prof. Fumitaka Kagawa (now in UTokyo), Profs. Naoki Ogawa, Xiuzhen Yu, and Dr. Kiyomi Nakajima in RIKEN CEMS; Prof. Atsushi Tsukazaki in Tohoku University; Dr. Naoki Shirakawa in AIST; Prof. Kazuhisa Kakurai and Dr. Takayasu Hanashima in CROSS.

I am grateful to Profs. Masashi Kawasaki, Youtarou Takahashi, Takahiro Morimoto, and Shuji Hasegawa for valuable comments and recommendations on this thesis. I also thank all the members of the Tokura Lab and of RIKEN for a lot of help and kindness.

Finally, I would like to express my special gratitude to my family for continuously giving me a lot of support and encouragement.

Contents

Chapter 1
Introduction

1.1 Topological Insulators

The notion of topology has given new ways to understand condensed-matter physics and led to discoveries of quantum phases including the topological insulators. The basic concept of topology is to classify geometrical objects by focusing on certain quantities preserved under some continuous deformations. For instance, as shown in Fig. 1.1, focusing on the number of holes (genus, g in mathematics) in objects, the mug and the doughnut (torus) can be deformed to each other without discontinuous manipulation, such as tearing somewhere. By contrast, the torus cannot be continuously deformed to a sphere because there is no hole. Thus, one can say that the torus and the mug are classified into the same topological class whereas the torus and the sphere belong to different classes.

By applying the notion of topology to gapped quantum material systems, the topological insulators, which broadly include the quantum Hall systems and time-reversal invariant topological insulators (TIs), are considered to be in a topologically non-trivial class (phase) of insulators distinct from normal (trivial) insulator phases. At the interface of TI and the trivial insulator (or a vacuum), a novel metallic edge or surface state emerges as a consequence of the discontinuous change of topology in two-dimensional (2D) or three-dimensional (3D) systems, respectively.

In this introductory section, we start with the review of the quantum Hall effect as a prototypical topological insulator in two-dimension with broken time-reversal symmetry, which provides the basic physics of TIs and the related topological quantum matter.

© The Author(s), under exclusive license to Springer Nature Singapore Pte Ltd. 2022
M. Mogi, *Quantized Phenomena of Transport and Magneto-Optics in Magnetic Topological Insulator Heterostructures*, Springer Theses,
https://doi.org/10.1007/978-981-19-2137-7_1

Fig. 1.1 Conceptual image of the example of topology in objects

1.1.1 Integer Quantum Hall Effect

When free electrons are subjected to an external magnetic field along z-direction, the motion of electrons has a circular orbit in x-y plane, called the cyclotron motion. Under a high magnetic field, the electrons confined to the circular orbit have quantized energy levels,

$$E_n = \hbar\omega_c \left(n + \frac{1}{2} \right) + \frac{\hbar^2 k_z^2}{2m}, \tag{1.1}$$

where ω_c is the cyclotron frequency and n is an integer. This energy level of $\hbar\omega_c\,(n + 1/2)$ is the Landau level and n is the filling factor.

When the electrons are confined to a two-dimensional (2D) (x-y) plane, the electrons have a full energy gap due to the Landau level formation, making the system insulating. However, on the contrary to the picture of the electronic bands, the system is still conductive due to the existence of gapless edge states. This phenomenon is the quantum Hall effect (QHE), of which the discovery is one of revolutionary advances in contemporary condensed-matter physics (Fig. 1.2) [1].

QHE was realized in 2D electron gas systems of Si-MOSFET and GaAs/AlGaAs heterostructures in the 1980s [1]. Regardless of the geometry of the samples as well as of the presence of inevitable disorders, the Hall resistance is precisely quantized to

$$R_{yx} = N \frac{h}{e^2}, \tag{1.2}$$

where h is Planck's constant, e is the elementary charge and N is an integer value. Simultaneously, the longitudinal resistance becomes zero ($R_{xx} = 0$), implying the dissipationless nature of the edge states of QHE.

Thouless, Kohmoto, Nightingale, and den Nijs (TKNN) have derived a formula to explain QHE from the Kubo formula [2]. According to the TKNN formula, Hall conductivity σ_{xy} is described by the energy bands below the Fermi energy E_F as follows:

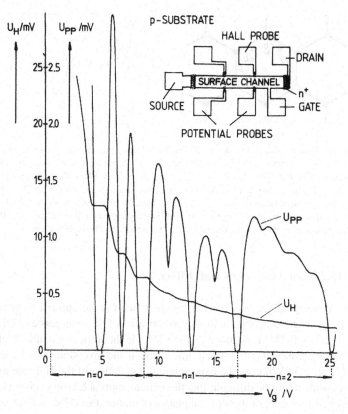

Fig. 1.2 Gate voltage dependence of longitudinal voltage (U_{pp}) and Hall voltage (U_H) in a Si-MOSFET device at $T = 1.5$ K and under $B = 18$ T. Reprinted figure with permission from [1] Copyright 1982 by the American Physical Society

$$\sigma_{xy} = \nu e^2 / h, \tag{1.3}$$

$$\nu = \sum_{n:\text{filled}} \int_{\text{BZ}} \frac{d^2 k}{2\pi} \left[\frac{\partial a_{n,y}}{\partial k_x} - \frac{\partial a_{n,x}}{\partial k_y} \right], \tag{1.4}$$

$$a_n(k) = -i \left\langle u_{nk} \left| \frac{\partial}{\partial k} \right| u_{nk} \right\rangle, \tag{1.5}$$

where $a_n(k)$ is the Berry connection characterized by the Bloch wave function $|u_{nk}\rangle$. ν is an integer value, which can be viewed as a topological invariant, the Chern number [3]. As a result, a 2D system in the QH state ($\nu \neq 0$) has conduction at its edge because it does not continuously connect with a vacuum or a trivial insulator ($\nu = 0$). Since the edge states originate from the cyclotron motion of electrons, they have chirality and conduct unidirectionally, avoiding backward scattering (i.e., $R_{xx} = 0$). The number of the chiral edge channels correspond to the Chern number ν calculated from the bulk properties, known as a bulk-boundary correspondence.

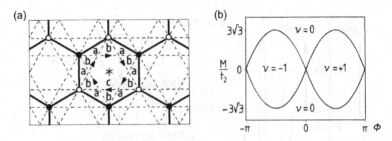

Fig. 1.3 **a** The Haldane model on a honeycomb lattice. The open and solid circles represent the *A* and *B* sublattice sites, respectively. The dashed lines represent the next-nearest-neighbor hopping, which experiences the magnetic flux through regions *a*, *b*, and *c*. **b** Phase diagram for the Haldane model. M, t_2, and ϕ represent the inversion-symmetry breaking on-site energy for *A* ($+M$) and *B* ($-M$) sites, the next-nearest-neighbor hopping term, and the phase which are acquired by the t_2 hopping, respectively. In the phase diagram, the QAH effect appears at $\nu = \pm 1$ regions. Reprinted figure with permission from [4] Copyright 1988 by the American Physical Society

1.1.1.1   Quantum Anomalous Hall Effect

The above topological aspects of Hall conductance can be applied to general 2D systems. Even at zero magnetic field, the possibilities of the emergence of QHE have been proposed by F. D. M. Haldane in 1988 [4]. He theoretically realized QHE in a honeycomb lattice model with broken time-reversal and inversion symmetries under zero field (Fig. 1.3). Whereas the experimental realization of the Haldane model in crystalline solids might be difficult, this theoretical approach is an exciting example for the recent discoveries of topological phases of matter; this effect is now called the quantum anomalous Hall (QAH) effect or Chern insulator. Since then, various theoretical examples for the realization of the QAH effect have been proposed in synthetic 2D systems with broken time-reversal symmetry [5–10]. Experimentally, in 2013, the QAH effect was discovered in a magnetically-doped TI, $Cr_x(Bi_{1-y}Sb_y)_{2-x}Te_3$ [11, 12] and, very recently, in intrinsic magnetic TI, $MnBi_2Te_4$ [13, 14] and twisted bilayer graphene [15].

1.1.2   Time-Reversal Invariant Topological Insulators

As described above, the Q(A)H insulators (Fig. 1.4a) are characterized by the Chern number under broken time-reversal symmetry conditions. On the other hand, time-reversal symmetric systems can be characterized by another topological invariant [16, 17]. While the Chern number is an integer number which corresponds to the number of chiral edge channels, the time-reversal invariant topological phases are classified by a Z_2 number (0 or 1), which corresponds to existence or absence of edge/surface modes for 2D/3D systems (Fig. 1.4b, c). The nontrivial Z_2 topology

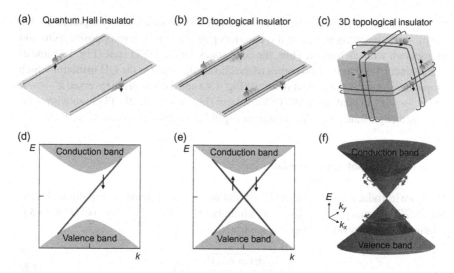

Fig. 1.4 **a–f** Schematics of the topological edge/surface modes and the electronic band structures for quantum Hall **a, d**, 2D topological (quantum spin Hall) **b, e**, and 3D topological insulators **c, f**

is guaranteed by the role of time-reversal symmetry for 1/2 spin for electrons as follows.

A time-reversal operator for particles with spin S is represented by $\mathcal{T} = \exp(i\pi S_y/\hbar)K$, where S_y is the y-component of the spin and K is complex conjugation. For spin 1/2 particles, $\mathcal{T}^2 = -1$, then leading to at least twofold degeneracies of all eigenstates for a time-reversal invariant Hamiltonian, which is Kramers' theorem [16, 17]. In a time-reversal invariant Bloch Hamiltonian, satisfying $\mathcal{T}\mathcal{H}(\boldsymbol{k})\mathcal{T}^{-1} = \mathcal{H}(-\boldsymbol{k})$, when $|\boldsymbol{k}, \boldsymbol{\sigma}\rangle$ is an eigenstate of the Hamiltonian, the time-reversal partner $\mathcal{T}|\boldsymbol{k}, \boldsymbol{\sigma}\rangle = |-\boldsymbol{k}, -\boldsymbol{\sigma}\rangle$ must also be an eigenstate of the Bloch Hamiltonian. When the spin up and down states are already degenerated, the Kramers pairs are degenerated at all the k points in the Brillouin zone. In the case of no spin degeneracy (i.e. in the presence of spin-orbit interaction), they are degenerated only at $\boldsymbol{k} = -\boldsymbol{k}$ (time-reversal invariant momentum, TRIM) points whereas they can split away from $\boldsymbol{k} = -\boldsymbol{k}$.

The Kramers theorem can classify the time-reversal invariant systems into two topologically different states. Let us consider two TRIM points, labeled $\Gamma_{a,b}$ (e.g., $\Gamma_a : (k_x, k_y) = (0, 0)$ and $\Gamma_b : (k_x, k_y) = (\pi/a, 0) = (-\pi/a, 0))$. We first focus on a Kramers pair at Γ_a. The two bands in a pair can split away from Γ_a by spin-orbit interaction. However, both the bands have to find partners at Γ_b. Here, there are two cases for their partners. One is that they again degenerate at Γ_b likewise at Γ_a and the other is that they switch their partners to other bands. Whereas the former case is allowed to have energy gaps, in the latter case, the bands must intersect E_F an odd number of times. This gapless state corresponds to the time-reversal invariant TI phase. For bulk insulating systems, the bands intersecting E_F eventually form at the edge or the surface. Furthermore, the edge/surface states are spin-polarized due to spin-orbit interaction, where the Kramers pairs of bands have opposite spins and momentum with each other. Thus, the edge states for the 2D TI state (Fig. 1.4b and e) can be understood as the two copies of the chiral edge state in the QH insulator which propagate along the opposite directions (Fig. 1.4a and d), which are reversed $k \rightarrow -k$ and $\sigma \rightarrow -\sigma$. In analogy to the QH insulator (Fig. 1.4a), the 2D TI phase with spin-polarized edge states is called a quantum spin Hall (QSH) insulator (Fig. 1.4b).

1.1.2.1 2D Topological Insulator

Here, we introduce a mathematical formulation for determining Z_2 topological numbers in 2D systems [17, 18]. By using a unitary matrix $w_{mn}(k) = \langle u_m(k)|\mathcal{T}|u_n(-k)\rangle$ relating the time-reversed Bloch wave functions of $|u_n(k)\rangle$,

$$\delta_i = \frac{\sqrt{\det[w(\Lambda_i)]}}{\text{Pf}[w(\Lambda_i)]} = \pm 1, \tag{1.6}$$

can be defined, where Λ_i is a TRIM point in the 2D Brillouin zone. By using δ_i, Z_2 topological invariant ν is given by

$$(-1)^\nu = \prod_{i=1}^{4} \delta_i. \tag{1.7}$$

When $\nu = 1$, the system is a 2D TI or a quantum spin Hall insulator.

If the system has inversion symmetry, Eq. 1.6 is greatly simplified to

$$\delta_i = \prod_{m=1}^{N} \xi_{2m}(\Gamma_i). \tag{1.8}$$

Here, $\xi_{2m}(\Gamma_i) = \pm 1$ is the parity eigenvalue of the $2m$th occupied energy band among $2N$ occupied bands at Γ_i, which shares the same parity eigenvalue $\xi_{2m} = \xi_{2m-1}$ with its Kramers degenerate partner [19]. By using this formulation, the 2D TI and trivial insulator phases can be determined by counting the number of 'band inversions' between the valence and conduction bands with different parity eigenvalues at each TRIM point.

Such a 2D TI phase has been discovered in HgTe/CdTe quantum wells [20, 21], InAs/GaSb quantum wells [22, 23], and a monolayer WTe$_2$ [24, 25], which exhibit quantum resistance due to nonlocal transport of the nondissipative edge states [26].

1.1.2.2 3D Topological Insulator

In 2007, the Z_2 topological characterization has been expanded to 3D systems, termed a 3D topological insulator (3D TI) [19, 27]. In a 3D TI, the Z_2 topological invariant has four components $(\nu_0; \nu_1\nu_2\nu_3)$ $(\nu_i = 0, 1 \ (i = 1, 2, 3))$. Following the discussion for the above 2D case, the mathematical formulation of ν_0 and $\nu_i = 0, 1 \ (i = 1, 2, 3)$ is given by,

$$(-1)^{\nu_0} = \prod_{i=1}^{4} \delta_i, \tag{1.9}$$

$$(-1)^{\nu_k} = \prod_{n_k=1; n_{j\neq k}=0,1} \delta_{i=(n_1 n_2 n_3)}. \tag{1.10}$$

The three indices of $(\nu_1\nu_2\nu_3)$, originating from the three-dimensionality, are characterized by the existence of the switching partner of Kramers pairs between TRIM points. When ν_1, ν_2, or ν_3 is 1, there is a closed Fermi surface in the surface Brillouin zone (i.e., surface state). More importantly, $\nu_0 = 1$ phase, called a strong TI, suggests the existence of an odd number of degeneracies at the Kramers points. By contrast, a condition of $\nu_0 = 0$ with ν_i $(i = 1, 2, 3)$ has an even number of Kramers points in a certain direction of the surface Brillouin zone, which is not robust against gap opening while keeping the Kramers degeneracy. Thus, $\nu_0 = 1$ phase is called a 'strong' TI because of the robustness against gap opening for an odd number of Kramers pairs while $\nu_0 = 0$ and $\nu_i \neq 0$ phase is called a 'weak' TI.

In the simplest case, the Hamiltonian for the surface state of the strong TI with a single Kramers point at the Γ point (the simplest case) as shown in Fig. 1.4f is described by

$$\mathcal{H} = \pm\hbar v_F(\sigma_x k_y - \sigma_y k_x). \tag{1.11}$$

This Hamiltonian is analogous to 2×2 Dirac Hamiltonian. While such a Dirac fermion state has also been found in graphene, the number of Dirac points (degenerated points at zero energy) is completely different from the 3D TI. In graphene, there are four Dirac points with opposite chiralities (2 valley × 2 spin), which is under the constraint of the fermion doubling theorem by Nielsen and Ninomiya [28, 29]. By contrast, a 3D TI can have only a single Dirac point. Instead, the two opposite flavors (or helicities for 3D TIs) of the surface Dirac fermions exist in the opposite surfaces, hence it does not violate the fermion doubling theorem. Such a single species of Dirac fermions with spin-polarization on the surface of 3D TIs produces exciting properties, such as highly efficient spintronic properties and exotic electromagnetic responses.

1.2   Exotic Electrodynamics of 3D Topological Insulator

In this section, we describe theoretical aspects of unique electromagnetic responses from the topological surface states, which are accessed by the generation of an energy gap by breaking time-reversal symmetry, such as applying an external magnetic field, doping a magnetic element, or proximity to a magnetic material.

1.2.1   Half-Integer Surface Quantum Hall Effect

Unlike conventional non-relativistic electrons, the Dirac electrons have exotic electronic properties, including delocalization transport [30, 31]. Especially, the Landau level formation and the QHE become unique. Because the zeroth Landau level locates at the charge neutrality of the Dirac dispersion, the Hall conductivity has an additional half-integer coefficient,

$$\sigma_{xy} = \left(n + \frac{1}{2}\right)\frac{e^2}{h}, \qquad (1.12)$$

where n is an integer value. In graphene, the four degeneracies of Dirac points result in the integer quantization of the Hall conductivity as $\sigma_{xy} = 4(n + 1/2)e^2/h$ [32–34]. In the case of a 3D TI, the top and bottom surface states will contribute to the total Hall conductivity of $\sigma_{xy} = (n_t + n_b + 1)e^2/h$, where $n_{t(b)}$ denote the highest filled Landau level index of the top (bottom) surface state [35].

Related to QHE of the Dirac electrons, the QAH effect, zero-field version of QHE, can be realized with magnetic element doping or proximity to a magnetic insulator through the exchange interaction. When the 2D Dirac electrons have a spin component normal to the 2D plane, the Dirac electrons gain a mass term m, corresponding to a gap at the Dirac point, so that the surface Hamiltonian becomes

$$\mathcal{H} = \pm \hbar v_F(\sigma_x k_y - \sigma_y k_x) + m\sigma_z \qquad (1.13)$$

$$= \boldsymbol{h} \cdot \boldsymbol{\sigma}, \qquad (1.14)$$

where $\boldsymbol{h} = (\pm \hbar v_F k_y, \mp \hbar v_F k_x, m)$. $\boldsymbol{h}(\boldsymbol{k})$ works as a fictitious field for the spin $\boldsymbol{\sigma}$. The unit vector $\hat{\boldsymbol{h}}(\boldsymbol{k}) = \boldsymbol{h}(\boldsymbol{k})/|\boldsymbol{h}(\boldsymbol{k})|$ has a winding number n wrapping a unit sphere,

$$n = \frac{1}{4\pi} \int d^2k \left(\frac{\partial \hat{\boldsymbol{h}}}{\partial k_x} \times \frac{\partial \hat{\boldsymbol{h}}}{\partial k_y}\right) \cdot \hat{\boldsymbol{h}} \qquad (1.15)$$

$$= \frac{1}{2}\frac{m}{|m|}, \qquad (1.16)$$

leading to the generation of a Berry's flux. According to the Berry's phase theory [36], the Hall conductivity is quantized to by $\sigma_{xy} = ne^2/h = \pm e^2/(2h)$ when the Fermi level locates within the gap.

Because a 3D TI can host a single Dirac fermion state, if we could measure only a single surface gapped with magnetism, we would measure the fractional (half-integer) QHE [37, 38]. The half-integer topology for a single species of Dirac fermions has been a fundamental problem in 2+1D quantum electrodynamics, called the 'parity anomaly' [39, 40]. Now, the 3D TI can be regarded as an excellent platform for the condensed-matter realization of the parity anomaly condensed matter [37, 38].

1.2.2 Axion Electrodynamics and Topological Magnetoelectric Effects

The TI surface state with the half-quantized Hall conductance is guaranteed by the 3D bulk properties. The low-energy electrodynamics for the 3D bulk state are well described in a Lagrangian form with a field theoretical approach [38].

$$\mathcal{L}_\theta = \theta \frac{e^2}{2\pi h} \boldsymbol{E} \cdot \boldsymbol{B}, \tag{1.17}$$

$$\theta = -\frac{1}{4\pi} \int_{BZ} d^3 k \epsilon_{ijk} \text{Tr}\left[A_i \partial_j A_k - i\frac{2}{3} A_i A_j A_k \right], \tag{1.18}$$

where BZ denotes the Brillouin zone, $A_i^{\mu\nu}$ is the Berry connection ($A_i^{\mu\nu} = \langle u^\mu | \partial k_i | u^\nu \rangle$) and $|u^\nu\rangle$ is the Bloch wave function of the filled band ν. The trace (Tr) sums over the occupied bands.

Focusing on the expression of Eq. 1.17, the response (effective action) is analogous to the axion electrodynamics [41]. The axion is a hypothetical particle in a theory of strong interaction (quantum chromodynamics, QCD) to explain the charge-parity (CP) symmetry in the QCD Lagrangian (strong CP problem) [42]. In the theory, θ appears as a CP symmetry violating term in the QCD Lagrangian. However, the experiment for the electric dipole moment of the neutron suggests that the CP symmetry is not violated, indicating that θ should be zero in the ground state. Peccei and Quinn have proposed the axion field to resolve the CP symmetry [43]. The axion particles are a kind of Nambu-Goldstone bosons, which are supposed to be generated by spontaneous symmetry breaking of CP symmetry [44, 45]. Currently, because the axions are highly expected to be a candidate of dark matter, much effort has been devoted to observing them experimentally [46].

In condensed-matter physics, θ can be finite in a wide class of crystalline insulators [47]. Especially, under time-reversal symmetry (i.e., for non-magnetic insulators), θ is limited to 0 or π, which corresponds to a trivial insulator or a 3D TI, respectively. The reason of this limitation is that the path integral $\exp[i \int d^3x dt \mathcal{L}_\theta] = e^{in\theta}$ (n is an integer value) gives the low-energy electrodynamic responses which are not changed by $\theta \to \theta + 2\pi$ assuming θ is constant in a closed space [48]. Under the time-reversal symmetry, the path integral, $\exp[i \int d^3x dt \mathcal{L}_\theta]$, is constant by the time-reversal operation ($\theta \to -\theta$). Thus, θ can be regarded as a topological invariant. At the interface between a TI and a trivial insulator, θ discontinuously changes, resulting in the emergence of the metallic interface state to cancel out the axionic response [49] that is consistent with the discussion of the Z_2 invariant derived by the topological aspects of the bulk wave functions. By breaking time-reversal symmetry, forming a gap in the surface state, θ is able to continuously change between 0 and 2π and now $\nabla\theta$ has a finite value.

We here describe the unusual electrodynamic responses caused by the axion θ term with broken time-reversal symmetry. By using the Eq. 1.17 in addition to the Lagrangian for electromagnetic fields $\mathcal{L}_{\text{EM}} = \epsilon_0 E^2/2 - B^2/(2\mu_0)$ (in SI units), Euler-Lagrange equation leads to Maxwell's equations modified by the θ term,

$$\nabla \cdot \boldsymbol{E} = \frac{e^2}{2\epsilon_0 h}\nabla\left(\frac{\theta}{\pi}\right)\cdot\boldsymbol{B}, \tag{1.19}$$

$$\nabla \cdot \boldsymbol{B} = 0, \tag{1.20}$$

$$\nabla \times \boldsymbol{B} = \frac{1}{c^2}\frac{\partial \boldsymbol{E}}{\partial t} + \frac{\mu_0 e^2}{2h}\nabla\left(\frac{\theta}{\pi}\right)\times\boldsymbol{E} + \frac{\mu_0 e^2}{2h}\frac{\dot{\theta}}{\pi}\boldsymbol{B}, \tag{1.21}$$

$$\nabla \times \boldsymbol{E} = -\frac{\partial \boldsymbol{B}}{\partial t}. \tag{1.22}$$

Thus, θ affects the electromagnetic responses only when θ has spatial ($\nabla\theta$) and/or time ($\partial_t\theta$) variations. Moreover, by using Maxwell's equation in matter, $\nabla \cdot \boldsymbol{D} = 0$, $\nabla \times \boldsymbol{H} - \dot{\boldsymbol{D}} = 0$, where $\boldsymbol{D}$ and $\boldsymbol{H}$ are the electric flux density and the magnetic field, respectively, and the relations with the electric polarization $\boldsymbol{P}$ and the magnetization $\boldsymbol{M}$, $\boldsymbol{D} = \epsilon_0\boldsymbol{E} + \boldsymbol{P}$, $\boldsymbol{H} = \boldsymbol{B}/\mu_0 - \boldsymbol{M}$,

$$\boldsymbol{P} = \frac{e^2}{2h}\frac{\theta}{\pi}\boldsymbol{B}, \tag{1.23}$$

$$\boldsymbol{M} = \frac{e^2}{2h}\frac{\theta}{\pi}\boldsymbol{E}, \tag{1.24}$$

are obtained. These relations indicate the magnetoelectric effect [50], where the magnetization is induced by an electric field (Fig. 1.5a) and the electric polarization is induced by a magnetic field (Fig. 1.5b). In a 3D TI, because of $\theta = \pi$ inside the bulk, the magnetoelectric effect may be quantized to a half-quantum value of $e^2/(2h)$ (topological magnetoelectric effect) is expected.

Besides, the axion electrodynamics can be related to the half-integer surface QHE as described in the previous section. When the Fermi level is within the gap, there is no dissipative current so that the current density flowing on the surface is described by the sum of the polarization current and the magnetization current, $\boldsymbol{j} = \partial_t\boldsymbol{P} + \nabla \times \boldsymbol{M} = (e^2/2h)\nabla(\theta/\pi) \times \boldsymbol{E}$ as derived from Eqs. 1.22, 1.23, 1.24.

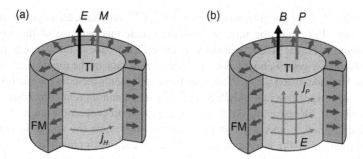

Fig. 1.5 a, b Magnetization induced by an electric field (**a**) and polarization induced by a magnetic field (**b**) in a TI coated by a ferromagnet

The half-quantized Hall current generated by an electric field would be observed by measuring the Faraday effect when the photon energy is much lower than the gap of the surface state, $E \ll E_g$ [38]. At the interface between a vacuum and TI, the electric field of light generates the Hall current, producing an additional electromagnetic field orthogonal to that of the incident light. This means that the boundary condition for the electromagnetic fields of the light is modified. Thus, the light polarization is rotated by an angle

$$\theta_F = \tan^{-1} \frac{\alpha}{1 + \sqrt{\epsilon/\mu}},\tag{1.25}$$

where α is the fine structure constant ($\sim 1/137$). ϵ and μ are the permittivity and permeability of the TI, respectively. In the simplest case of $\epsilon = \mu = 1$, $\theta_F \simeq \alpha \simeq 7.3 \times 10^{-3}$ rad.

One other intriguing response derived from the axion electrodynamics is the emergence of image magnetic monopoles induced by placing a point-like charge on a TI surface. Rewriting Eq. 1.21 as $\nabla \times \boldsymbol{B} = (\alpha/c)\hat{n} \times \boldsymbol{E}$, of which the right-hand side of the equation corresponds to the surface QH current density of $\boldsymbol{j} = (e^2/2h)\nabla(\theta/\pi) \times \boldsymbol{E}$. Thus, a magnetic field from the TI surface is produced, being consistent with the existence of a magnetic monopole inside the TI. Such a magneto-electric composite particle can be viewed as a 'dyon' in terms of high-energy physics [51], carrying fractional statistics thanks to the half-integer quantized axion electrodynamics.

1.3 3D Topological Insulator Materials

Hereafter, we focus on the theoretical search and the experimental realization of 3D TI materials. To realize the exotic electrodynamics of 3D TIs as introduced in the previous section, highly insulating bulk states and precise Fermi level control are essential. Among the currently discovered 3D TI materials, tetradymite Bi_2Se_3, Bi_2Te_3, and Sb_2Te_3 are the most well-studied materials because of the simplest surface band structure characterized by a single Dirac cone at the Γ point and the large bulk energy gap of about 0.3 eV [52]. The existence of the topological surface states has been identified with surface sensitive experiments including angle-resolved photoemission spectroscopy (ARPES) [53–57] and scanning tunneling microscopy (STM) [58–60]. Moreover, to verify the surface transport properties, tuning E_F close to the Dirac point is necessary. In this section, we review the research works of the confirmation of the surface states and the recent development of 3D TI bulk crystal/film growth, which are crucial for the realization of applicable electrical properties emerged from the topological surface states.

1.3.1 Single Dirac Fermion Surface States in Bi_2Se_3, Sb_2Te_3, and Bi_2Te_3

In 2008, the 3D TI phase was firstly discovered in a $Bi_{1-x}Sb_x$ alloy [19, 53, 54]. Soon after the discovery, Bi_2Se_3, Sb_2Te_3, and Bi_2Te_3 were also confirmed to be 3D TIs [52, 55–57]. While $Bi_{1-x}Sb_x$ has a smaller bulk gap ($\sim$0.05 eV) and a complex surface band structure, where E_F crosses five branches of the surface bands, Bi_2Se_3, Sb_2Te_3, and Bi_2Te_3 have a large bulk gap ($\sim$0.3 eV) and a single Dirac cone. Therefore, by tuning E_F, we can enjoy the 2D Dirac physics on the surface. Furthermore, they have layered crystal structures, where a monolayer has a Se(Te)-Bi(Sb)-Se(Te)-Bi(Sb)-Se(Te) stacking sequence along c axis (1 quintuple layer (QL) $\sim$1 nm in thickness) and the monolayers are bonded by van der Waals force. Such layered structures are beneficial for surface sensitive experiments because shiny cleavage surfaces are easily obtained.

Theoretically, the nontrivial Z_2 topology and the resultant topological surface bands of Bi_2Se_3, Sb_2Te_3, and Bi_2Te_3 are verified by calculating Z_2 topological invariant and first-principles calculation in slab models, respectively. Owing to the centrosymmetric crystal structure of Bi_2Se_3, Sb_2Te_3, and Bi_2Te_3, the Z_2 topological invariant can be calculated via investigating the parity eigenvalues of the wave functions only at TRIM points of the Brillouin zone [37]. At the Γ point of Bi_2Se_3, the valence and conduction bands come from Se $4p$ orbitals and Bi $6p$ orbitals when the spin-orbit coupling is ignored. Turning on the spin-orbit coupling, the highest Se $4p$ valence band and the lowest Bi $6p$ conduction band with opposite parity eigenvalues are inverted at the Γ point while kept at the other TRIM points. Thus, Bi_2Se_3, as well as Sb_2Te_3 and Bi_2Te_3, are classified into a (strong) 3D TI phase. Indeed, first-principles calculation shows a single Dirac cone on each surface, being consistent with the discussion of Z_2 topological invariant.

To experimentally confirm the existence of the surface state, surface-sensitive ARPES have been employed [59, 60]. The ARPES mapping clearly show the surface bands characterized by linearly dispersive Dirac cones. Furthermore, by using spin-resolved ARPES, the spin-momentum locking, where the spin polarization is perpendicular to the momentum, was also confirmed.

STM can probe the modification of the electronic structures under external magnetic fields. In contrast to the Landau level splitting (Eq. 1.1) for conventional non-relativistic electrons with an energy-momentum relation of $E \propto k^2$, the energy E_n of n-th Landau level for 2D Dirac fermions is expressed as

$$E_n = v_\mathrm{F}\sqrt{2e\hbar|n|B}. \tag{1.26}$$

Therefore, E_n is proportional to $\sqrt{nB}$ for the 2D Dirac cones while it is proportional to nB for the conventional electrons. STM provides the spectroscopic evidence of the $\sqrt{nB}$ Landau level splitting. By measuring the tunneling conductance between the sample and the STM tip under magnetic fields, we can observe peak structures in the density of states due to the Landau level formation. With increasing the magnetic fields, the intervals of the peaks become large. By plotting the energy levels of the peaks as a function of $\sqrt{nB}$, they actually scale with $\sqrt{nB}$ [59].

1.3.2 3D TIs with Excellent Transport Properties

Whereas the existence of the topological surface state has been established in the above 3D TI crystals by surface sensitive experiments, many unique surface transport properties have been still elusive because the residual bulk carriers hinder them. Key issues are to reduce the chemical defects and to control the Fermi energy. Chemistry and device technology make them possible.

1.3.2.1 Bulk Crystals

As demonstrated in ARPES studies, Bi_2Se_3, Sb_2Te_3, and Bi_2Te_3 have naturally occurring defects and then the carriers are unintentionally doped. To reduce the defects, Bi_2Te_2Se has been synthesized, where the ordered Te-Bi-Se-Bi-Te QLs are formed and the stoichiometry is well-controlled [61]. Furthermore, to control E_F, $Bi_{2-x}Sb_xTe_{3-y}Se_y$ has been proposed and grown [62]. Indeed, by tuning the composition x and y, E_F can be tuned close to the Dirac point.

A smoking gun experiment to establish the surface-dominated transport is the observation of QHE arising from the surface states. Exfoliation of $Bi_{2-x}Sb_xTe_{3-y}Se_y$ flakes down to hundreds of nm can achieve further decrease of residual bulk carriers. Then, QHE was observed under an external magnetic field. Interestingly, by gating with the SiO_2/Si substrate below the $Bi_{2-x}Sb_xTe_{3-y}Se_y$ flake, the carrier density of the bottom surface state can be tuned. Then, the coefficient of the quantized Hall conductivity is changed by an integer quantized value. This integer quantization can be understood by the sum of the Hall conductivity for the top surface ($\sigma_{xy}^{t} = (\nu_t + 1/2)\, e^2/h$) and bottom surface ($\sigma_{xy}^{b} = (\nu_b + 1/2)\, e^2/h$) states,

$$\sigma_{xy} = \sigma_{xy}^{t} + \sigma_{xy}^{b} = (\nu_t + \nu_b + 1)\,\frac{e^2}{h}, \qquad (1.27)$$

where the subscripts of t and b stand for the top and bottom surface states, respectively.

More recently, through the chemical engineering of the defects, a Sn-doped $Bi_{1.1}Sb_{0.9}Te_2S$ bulk crystal has been made, which has even higher bulk insulating transport properties than previously reported 3D TI crystals [63]. The surface QHE has also been observed in similar compositions of flakes [64, 65].

1.3.2.2 Thin Films

Thin-film growth is also an effective way to fabricate bulk insulating 3D TIs. One reason is reducing the bulk region by precisely thinning the crystals, realizing the surface-dominated transport. One other reason is the capability of fine tuning of composition, which enables us to precisely control the Fermi level. One such ideal example for the highly tunable 3D TI thin film is (Bi, Sb)$_2$Te$_3$ [66]. Since Bi$_2$Te$_3$ has electron-type bulk carriers (induced by Te vacancies) and Sb$_2$Te$_3$ has hole-type bulk carriers (induced by Sb-Te anti-site defects), mixing them would lead to the charge compensation. Indeed, ARPES experiments demonstrate the successful E_F tuning in molecular-beam epitaxy (MBE) grown (Bi, Sb)$_2$Te$_3$. Since the Dirac point is isolated from the bulk band, this E_F tuning can access the physics related to the zero-energy Dirac fermions.

More precise E_F tuning is accomplished by the fabrication of a field effect transistor (FET) device. Indeed, the QHE is observed in a (Bi, Sb)$_2$Te$_3$ FET device [67]. By applying gate voltage, $\nu = \pm 1$ QH plateaus were observed, being consistent with Eq. 1.27. In between the $\nu = \pm 1$ QH plateaus, the enhancement of R_{xx} indicates that E_F is tuned close to the Dirac point.

1.3.3 Magnetically Doped Topological Insulators

Breaking time-reversal symmetry of 3D TIs allows us to realize exotic electrodynamics of 3D TI including the QAH effect. The time-reversal symmetry breaking opens a gap at the Dirac point of the surface state and makes the surface state insulating when E_F resides in the gap. The original idea to break time-reversal symmetry is proximitizing a magnetic insulator. One other successful way is doping magnetic ions [11], such as Mn, Fe, Cr, and V, in analogy with diluted magnetic semiconductors [68]. Here, we briefly review the studies of the magnetically doped 3D TIs, which lead to the first observation of the QAH effect.

1.3.3.1 Massive Dirac Fermions on the Surface of TI

To achieve the most interesting regime of 3D TIs with broken time-reversal symmetry, long-range magnetic ordering should be realized in the insulating state (E_F lies within the gap). In contrast to the diluted magnetic semiconductors with carrier-mediated ferromagnetic ordering (such as the Ruderman-Kittel-Kasuya-Yosida (RKKY) interaction) [68], 3D TIs doped with magnetic impurities can have a long-range magnetic order without carriers [11]. Such exchange interaction between TI bulk states and the localized spin states of the magnetic impurities is driven by the virtual interband transition through the localized spin states, so-called the van Vleck mechanism or the Bloembergen-Rowland interaction [69]. Especially, this mechanism is enhanced due to the features that wavefunctions for both the valence and conduction bands have a similar character to each other, in that they are p-like and also their bands are partially inverted due to spin-orbit coupling [11].

When the long-range ferromagnetic order is realized and the magnetization points to the surface normal $\hat{z}$, the surface Hamiltonian expressed by a single Dirac cone becomes

$$\mathcal{H} = \hbar v_F (k_x \sigma_y - k_y \sigma_x) - J n S_z \sigma_z, \tag{1.28}$$

where J is the exchange coupling constant between the surface state and localized spin-polarized states of the magnetic impurities, n is the average areal density of magnetic dopants, and S_z is the z-component of the electrons' spin in the magnetic impurities. Then, the size of the magnetically induced gap Δ amounts to $\Delta = J M_z / \mu_B$, where $M_z = n S_z \mu_B$ is the magnetization normal to the surface and μ_B is the Bohr magneton.

ARPES studies have found a sizeable gap in the magnetically doped TIs [70–72]. The gap size was estimated to be $2\Delta \sim 50$ meV [70]. Likewise, STM experiments observe the magnetization gap of about $2\Delta \sim 70$ meV [73]. Furthermore, STM can clarify the spatial distribution of the magnetization gap size, which actually has large spatial inhomogeneity This spatial variation of the gap size is strongly correlated with the spatial inhomogeneity of the distribution of the magnetic (Cr) ions [73].

1.3.3.2 Observation of Quantum Anomalous Hall Effect

In 2013, Chang et al. firstly observed the quantum anomalous Hall (QAH) effect in a 5-QL Cr-doped (Bi, Sb)$_2$Te$_3$ thin film at a temperature of 30 mK [12]. By doping Cr ions, the 3D TI shows the long-range ferromagnetic order with the Curie temperature (T_C) of about 15 K. Fine tuning of Bi/Sb ratio and back gating lead to the precise E_F control [74]. By tuning the gate voltage, the quantized Hall resistance ($\rho_{yx} = h/e^2$) accompanying almost zero ρ_{xx} is observed even at zero field. During the magnetic reversal at around the coercive field (H_c), the sign of ρ_{yx} changes, indicating the direction of chiral edge states is reversed. At around $H = H_c$, ρ_{xx} increases, which is understood by the interference between chiral edge states formed on the magnetic domain walls [75].

This observation has currently been reproduced in many groups [76–79]. They discuss more details of transport properties, such as the universality of QHE and the QAH effect, the precision of quantization with thermal activation behaviors [79, 80], magnetic dynamics [81], etc. Furthermore, V-doped (Bi, Sb)$_2$Te$_3$ also exhibits the QAH effect with a large coercive field ($\mu_0 H_c \sim 1$ T) [82, 83]. However, the observable temperature is still been below 100 mK even though the size of the magnetization gap is expected to be several tens of meV. The inhomogeneity of the magnetization gap and the formation of in-gap states [84, 85] might be one of the reasons for the low stability of the QAH state [73].

1.4   Purpose of Thesis

In this thesis, we explore quantized phenomena in transport and magneto-optical responses in magnetic TIs. While the observation of the QAH effect in a magnetic TI thin film was a paradigmatic advance in this research field, increasing its stability (or increasing the observable temperature) and studying advanced physics have been impeded due to inhomogeneous distributions of structural, charge and magnetic disorders as partly probed by the microscopic measurements. Here, we design heterostructure based on the magnetic TIs by using molecular-beam epitaxy thin-film growth technique. For instance, via the development of a modulation doping technique of magnetic elements and the synthesis of heterostructures with ferromagnetic insulators, we explore the formation of "clean" magnetic TI systems. Furthermore, magnetic controls in such heterostructures enable us to realize an axion insulator state potentially exhibiting the quantized magnetoelectric effect and also to observe the half-integer Hall conductance associated with the parity anomaly of a single species of Dirac fermions on the surface of TI. The constitution of this thesis is as follows.

In Chap. 2, we introduce experimental techniques and methods used in the works of this thesis.

In Chap. 3, we develop a magnetic modulation doping method in Cr-doped (Bi, Sb)$_2$Te$_3$ thin films. Compared to the conventional uniform doping, the magnetic elements are intensively doped only in the vicinity of the surface, leading to strong exchange coupling between the surface states and the magnetic elements. We achieve an increased observable temperature of the QAH effect up to 2 K from the previously reported observable temperature as low as 50 mK.

In Chap. 4, we study magnetic proximity effect on TI surface state. Instead of magnetic doping, by proximitizing ferromagnetic insulators, the topological surface states acquire a magnetization gap through interfacial exchange coupling. In this study, we grew Te-based van der Waals ferromagnetic insulators, Cr$_2$Ge$_2$Te$_6$ and Cr$_2$Si$_2$Te$_6$ and the heterostructures with (Bi, Sb)$_2$Te$_3$. We find efficient exchange coupling at their interfaces, realizing the QAH effect and also the current-induced switching utilizing the spin-momentum locking of the TI surface states.

In Chap. 5, we explore topological phase transitions relevant to the QAH effect. First, we study a QAH insulator - trivial insulator transition upon the rotation of the samples under an external magnetic field. This rotates the magnetization, leading to the competition between the magnetization-induced gap and the surface-hybridization induced trivial gap. Furthermore, we explore a new quantum phase termed as an axion insulator phase in the synthetic magnetic heterostructures with surface-selective spin degrees of freedom in the 3D TI as developed in Chap. 3.

In Chap. 6, we study exotic electrodynamics associated with the single Dirac fermion state on the surface of TI. In general 2D systems, two flavors of Dirac fermions must exist in momentum space, which is known as the Nielsen-Ninomiya fermion doubling theorem. On the other hand, in a 3D TI, only one flavor of Dirac fermions can exist on each surface. Hence, the 3D TI is a fascinating system to study the single Dirac fermions physics. One important property of the single Dirac fermion state is half-integer quantized Hall current generation under an electric field. Such Hall current generation is known as a consequence of the parity anomaly of 2+1D quantum electrodynamics. Here, we observe the half-quantization phenomena in a synthetic 'semi-magnetic' TI through terahertz spectroscopy and electrical transport studies. We establish the half-quantized phenomena and the realization of the parity anomaly in condensed-matter systems.

Finally, we summarize this thesis in Chap. 7.

References

1. K.v. Klitzing, G. Dorda, M. Pepper, Phys. Rev. Lett. **45**(6), 494 (1980)
2. D.J. Thouless, M. Kohmoto, M.P. Nightingale, M. den Nijs, Phys. Rev. Lett. **49**(6), 405 (1982)
3. M. Kohmoto, Ann. Phys. **160**(2), 343 (1985)
4. F.D.M. Haldane, Phys. Rev. Lett. **61**(18), 2015 (1988)
5. M. Onoda, N. Nagaosa, Phys. Rev. Lett. **90**(20), 206601 (2003)
6. C.X. Liu, X.L. Qi, X. Dai, Z. Fang, S.C. Zhang, Phys. Rev. Lett. **101**(14), 146802 (2008)
7. Z. Qiao, S.A. Yang, W. Feng, W.K. Tse, J. Ding, Y. Yao, J. Wang, Q. Niu, Phys. Rev. B **82**(16), 161414 (2010)
8. G. Xu, H. Weng, Z. Wang, X. Dai, Z. Fang, Phys. Rev. Lett. **107**(18), 186806 (2011)
9. D. Xiao, W. Zhu, Y. Ran, N. Nagaosa, S. Okamoto, Nat. Commun. **2**, 596 (2011)
10. H. Zhang, J. Wang, G. Xu, Y. Xu, S.C. Zhang, Phys. Rev. Lett. **112**(9), 096804 (2014)
11. R. Yu, W. Zhang, H.J. Zhang, S.C. Zhang, X. Dai, Z. Fang, Science **329**(5987), 61 (2010)
12. C.Z. Chang, J. Zhang, X. Feng, J. Shen, Z. Zhang, M. Guo, K. Li, Y. Ou, P. Wei, L.L. Wang et al., Science **340**(6129), 167 (2013)
13. Y. Deng, Y. Yu, M.Z. Shi, J. Wang, X.H. Chen, Y. Zhang, arXiv preprint arXiv:1904.11468 (2019)
14. C. Liu, Y. Wang, H. Li, Y. Wu, Y. Li, J. Li, K. He, Y. Xu, J. Zhang, Y. Wang, arXiv preprint arXiv:1905.00715 (2019)
15. M. Serlin, C. Tschirhart, H. Polshyn, Y. Zhang, J. Zhu, K. Watanabe, T. Taniguchi, L. Balents, A. Young, Science (2019)
16. C.L. Kane, E.J. Mele, Phys. Rev. Lett. **95**(22), 226801 (2005)
17. C.L. Kane, E.J. Mele, Phys. Rev. Lett. **95**(14), 146802 (2005)
18. L. Fu, C.L. Kane, Phys. Rev. B **74**(19), 195312 (2006)
19. L. Fu, C.L. Kane, E.J. Mele, Phys. Rev. Lett. **98**(10), 106803 (2007)

20. B.A. Bernevig, T.L. Hughes, S.C. Zhang, Science **314**(5806), 1757 (2006)
21. M. König, S. Wiedmann, C. Brüne, A. Roth, H. Buhmann, L.W. Molenkamp, X.L. Qi, S.C. Zhang, Science **318**(5851), 766 (2007)
22. C. Liu, T.L. Hughes, X.L. Qi, K. Wang, S.C. Zhang, Phys. Rev. Lett. **100**(23), 236601 (2008)
23. I. Knez, R.R. Du, G. Sullivan, Phys. Rev. Lett. **107**(13), 136603 (2011)
24. X. Qian, J. Liu, L. Fu, J. Li, Science **346**(6215), 1344 (2014)
25. S. Wu, V. Fatemi, Q.D. Gibson, K. Watanabe, T. Taniguchi, R.J. Cava, P. Jarillo-Herrero, Science **359**(6371), 76 (2018)
26. A. Roth, C. Brüne, H. Buhmann, L.W. Molenkamp, J. Maciejko, X.L. Qi, S.C. Zhang, Science **325**(5938), 294 (2009)
27. J.E. Moore, L. Balents, Phys. Rev. B **75**(12), 121306 (2007)
28. H.B. Nielsen, M. Ninomiya, Nucl. Phys. B **185**(1), 20 (1981)
29. H.B. Nielsen, M. Ninomiya, Nucl. Phys. B **193**(1), 173 (1981)
30. K. Ziegler, Phys. Rev. Lett. **80**, 3113 (1998). https://doi.org/10.1103/PhysRevLett.80.3113. https://link.aps.org/doi/10.1103/PhysRevLett.80.3113
31. K. Nomura, M. Koshino, S. Ryu, Phys. Rev. Lett. **99**(14), 146806 (2007)
32. Y. Zheng, T. Ando, Phys. Rev. B **65**, 245420 (2002). https://doi.org/10.1103/PhysRevB.65.245420. https://link.aps.org/doi/10.1103/PhysRevB.65.245420
33. K.S. Novoselov, A.K. Geim, S.V. Morozov, D. Jiang, M.I. Katsnelson, I.V. Grigorieva, S. Dubonos, A.A. Firsov, Nature **438**(7065), 197 (2005)
34. Y. Zhang, Y.W. Tan, H.L. Stormer, P. Kim, Nature **438**(7065), 201 (2005)
35. D.H. Lee, Phys. Rev. Lett. **103**, 196804 (2009). https://doi.org/10.1103/PhysRevLett.103.196804. https://link.aps.org/doi/10.1103/PhysRevLett.103.196804
36. N. Nagaosa, J. Sinova, S. Onoda, A.H. MacDonald, N.P. Ong, Rev. Mod. Phys. **82**(2), 1539 (2010)
37. L. Fu, C.L. Kane, Phys. Rev. B **76**(4), 045302 (2007)
38. X.L. Qi, T.L. Hughes, S.C. Zhang, Phys. Rev. B **78**(19), 195424 (2008)
39. A.J. Niemi, G.W. Semenoff, Phys. Rev. Lett. **51**(23), 2077 (1983)
40. R. Jackiw, Phys. Rev. D **29**(10), 2375 (1984)
41. F. Wilczek, Phys. Rev. Lett. **58**(18), 1799 (1987)
42. R.D. Peccei, in *Axions* (Springer, Berlin, 2008), pp. 3–17
43. R.D. Peccei, H.R. Quinn, Phys. Rev. Lett. **38**(25), 1440 (1977)
44. S. Weinberg, Phys. Rev. Lett. **40**(4), 223 (1978)
45. F. Wilczek, Phys. Rev. Lett. **40**(5), 279 (1978)
46. J.E. Kim, G. Carosi, Rev. Mod. Phys. **82**(1), 557 (2010)
47. A.M. Essin, J.E. Moore, D. Vanderbilt, Phys. Rev. Lett. **102**(14), 146805 (2009)
48. E. Witten, Rev. Mod. Phys. **88**(3), 035001 (2016)
49. S. Coh, D. Vanderbilt, A. Malashevich, I. Souza, Phys. Rev. B **83**, 085108 (2011). https://doi.org/10.1103/PhysRevB.83.085108. https://link.aps.org/doi/10.1103/PhysRevB.83.085108
50. M. Fiebig, J. Phys. D **38**(8), R123 (2005)
51. E. Witten, Phys. Lett. B **86**(3–4), 283 (1979)
52. H. Zhang, C.X. Liu, X.L. Qi, X. Dai, Z. Fang, S.C. Zhang, Nat. Phys. **5**(6), 438 (2009)
53. D. Hsieh, D. Qian, L. Wray, Y. Xia, Y.S. Hor, R.J. Cava, M.Z. Hasan, Nature **452**(7190), 970 (2008)
54. D. Hsieh, Y. Xia, L. Wray, D. Qian, A. Pal, J. Dil, J. Osterwalder, F. Meier, G. Bihlmayer, C. Kane et al., Science **323**(5916), 919 (2009)
55. Y. Xia, D. Qian, D. Hsieh, L. Wray, A. Pal, H. Lin, A. Bansil, D. Grauer, Y.S. Hor, R.J. Cava et al., Nat. Phys. **5**(6), 398 (2009)
56. Y. Chen, J.G. Analytis, J.H. Chu, Z. Liu, S.K. Mo, X.L. Qi, H. Zhang, D. Lu, X. Dai, Z. Fang et al., Science **325**(5937), 178 (2009)
57. D. Hsieh, Y. Xia, D. Qian, L. Wray, J. Dil, F. Meier, J. Osterwalder, L. Patthey, J. Checkelsky, N.P. Ong et al., Nature **460**(7259), 1101 (2009)
58. P. Roushan, J. Seo, C.V. Parker, Y.S. Hor, D. Hsieh, D. Qian, A. Richardella, M.Z. Hasan, R.J. Cava, A. Yazdani, Nature **460**(7259), 1106 (2009)

59. T. Hanaguri, K. Igarashi, M. Kawamura, H. Takagi, T. Sasagawa, Phys. Rev. B **82**(8), 081305 (2010)
60. P. Cheng, C. Song, T. Zhang, Y. Zhang, Y. Wang, J.F. Jia, J. Wang, Y. Wang, B.F. Zhu, X. Chen et al., Phys. Rev. Lett. **105**(7), 076801 (2010)
61. Z. Ren, A. Taskin, S. Sasaki, K. Segawa, Y. Ando, Phys. Rev. B **82**(24), 241306 (2010)
62. A.A. Taskin, Z. Ren, S. Sasaki, K. Segawa, Y. Ando, Phys. Rev. Lett. **107**, 016801 (2011). https://doi.org/10.1103/PhysRevLett.107.016801. https://link.aps.org/doi/10.1103/PhysRevLett.107.016801
63. S. Kushwaha, I. Pletikosić, T. Liang, A. Gyenis, S. Lapidus, Y. Tian, H. Zhao, K. Burch, J. Lin, W. Wang et al., Nat. Commun. **7**, 11456 (2016)
64. F. Xie, S. Zhang, Q. Liu, C. Xi, T.T. Kang, R. Wang, B. Wei, X.C. Pan, M. Zhang, F. Fei, X. Wang, L. Pi, G.L. Yu, B. Wang, F. Song, Phys. Rev. B **99**, 081113 (2019). https://doi.org/10.1103/PhysRevB.99.081113. https://link.aps.org/doi/10.1103/PhysRevB.99.081113
65. K. Ichimura, S.Y. Matsushita, K.K. Huynh, K. Tanigaki, Appl. Phys. Lett. **115**(5), 052104 (2019)
66. J. Zhang, C.Z. Chang, Z. Zhang, J. Wen, X. Feng, K. Li, M. Liu, K. He, L. Wang, X. Chen et al., Nat. Commun. **2**, 574 (2011)
67. R. Yoshimi, A. Tsukazaki, Y. Kozuka, J. Falson, K. Takahashi, J. Checkelsky, N. Nagaosa, M. Kawasaki, Y. Tokura, Nat. Commun. **6**, 6627 (2015)
68. T. Dietl, H. Ohno, Rev. Mod. Phys. **86**(1), 187 (2014)
69. N. Bloembergen, T.J. Rowland, Phys. Rev. **97**, 1679 (1955). https://doi.org/10.1103/PhysRev.97.1679. https://link.aps.org/doi/10.1103/PhysRev.97.1679
70. Y. Chen, J.H. Chu, J. Analytis, Z. Liu, K. Igarashi, H.H. Kuo, X. Qi, S.K. Mo, R. Moore, D. Lu et al., Science **329**(5992), 659 (2010)
71. L.A. Wray, S.Y. Xu, Y. Xia, D. Hsieh, A.V. Fedorov, Y. San Hor, R.J. Cava, A. Bansil, H. Lin, M.Z. Hasan, Nat. Phys. **7**(1), 32 (2011)
72. S.Y. Xu, M. Neupane, C. Liu, D. Zhang, A. Richardella, L.A. Wray, N. Alidoust, M. Leandersson, T. Balasubramanian, J. Sánchez-Barriga et al., Nat. Phys. **8**(8), 616 (2012)
73. I. Lee, C.K. Kim, J. Lee, S.J. Billinge, R. Zhong, J.A. Schneeloch, T. Liu, T. Valla, J.M. Tranquada, G. Gu et al., Proc. Natl. Acad. Sci. U.S.A. **112**(5), 1316 (2015)
74. C.Z. Chang, J. Zhang, M. Liu, Z. Zhang, X. Feng, K. Li, L.L. Wang, X. Chen, X. Dai, Z. Fang et al., Adv. Mater. **25**(7), 1065 (2013)
75. J. Wang, B. Lian, S.C. Zhang, Phys. Rev. B **89**, 085106 (2014). https://doi.org/10.1103/PhysRevB.89.085106. https://link.aps.org/doi/10.1103/PhysRevB.89.085106
76. J. Checkelsky, R. Yoshimi, A. Tsukazaki, K. Takahashi, Y. Kozuka, J. Falson, M. Kawasaki, Y. Tokura, Nat. Phys. **10**, 731 (2014). https://doi.org/10.1038/nphys3053
77. X. Kou, S.T. Guo, Y. Fan, L. Pan, M. Lang, Y. Jiang, Q. Shao, T. Nie, K. Murata, J. Tang, Y. Wang, L. He, T.K. Lee, W.L. Lee, K.L. Wang, Phys. Rev. Lett. **113**, 137201 (2014). https://doi.org/10.1103/PhysRevLett.113.137201. https://link.aps.org/doi/10.1103/PhysRevLett.113.137201
78. A. Kandala, A. Richardella, S. Kempinger, C.X. Liu, N. Samarth, Nat. Commun. **6**, 7434 (2015). https://doi.org/10.1038/ncomms8434
79. A. Bestwick, E. Fox, X. Kou, L. Pan, K.L. Wang, D. Goldhaber-Gordon, Phys. Rev. Lett. **114**(18), 187201 (2015)
80. E.J. Fox, I.T. Rosen, Y. Yang, G.R. Jones, R.E. Elmquist, X. Kou, L. Pan, K.L. Wang, D. Goldhaber-Gordon, Phys. Rev. B **98**, 075145 (2018). https://doi.org/10.1103/PhysRevB.98.075145. https://link.aps.org/doi/10.1103/PhysRevB.98.075145
81. E.O. Lachman, A.F. Young, A. Richardella, J. Cuppens, H. Naren, Y. Anahory, A.Y. Meltzer, A. Kandala, S. Kempinger, Y. Myasoedov et al., Sci. Adv. **1**(10), e1500740 (2015)
82. C.Z. Chang, W. Zhao, D.Y. Kim, H. Zhang, B.A. Assaf, D. Heiman, S.C. Zhang, C. Liu, M.H. Chan, J.S. Moodera, Nat. Mater. **14**(5), 473 (2015)
83. S. Grauer, S. Schreyeck, M. Winnerlein, K. Brunner, C. Gould, L. Molenkamp, Phys. Rev. B **92**(20), 201304 (2015)
84. Z. Yue, M. Raikh, Phys. Rev. B **94**(15), 155313 (2016)
85. M. Kawamura, R. Yoshimi, A. Tsukazaki, K.S. Takahashi, M. Kawasaki, Y. Tokura, Phys. Rev. Lett. **119**(1), 016803 (2017)

Chapter 2
Experimental Methods

2.1 Thin Film Growth

Molecular beam epitaxy (MBE) is a powerful method to fabricate thin films with finely tuned thickness and compositions and to design heterostructures with sharp interfaces. By evaporating elements from sources introduced in Knudsen cells (K-cells) under a high-vacuum condition, the gaseous elements crystallize on heated substrates. Our MBE system (purchased from EIKO Engineering) as shown in Fig. 2.1 consists of a load-lock chamber ($\sim 1 \times 10^{-5}$ Pa) and a growth chamber ($\sim 1 \times 10^{-7}$ Pa) equipped with five K-cells and RHEED (reflection high energy electron diffraction) system (kSA 400, k-Space). We grew 3D TI (Bi, Sb)$_2$Te$_3$, magnetically doped TI Cr$_x$(Bi$_{1-y}$Sb$_y$)$_{2-x}$Te$_3$ and V$_x$(Bi$_{1-y}$Sb$_y$)$_{2-x}$Te$_3$, and ferromagnetic insulators Cr$_2$Ge$_2$Te$_6$ and Cr$_2$Si$_2$Te$_6$ thin films on commercially available epi-ready semi-insulating InP(111)A (In termination) substrates. The two-inches InP substrates were cut into ~ 5 mm $\times$ 10 mm and fixed on a substrate holder made of stainless steel by clamping with steel plates with screws. Before loading the substrate holder into the load-lock chamber, we blow off the dust on the substrate surface with nitrogen gas. When the substrates were transferred to the growth chamber, they were pre-annealed up to 380 °C to further clean the surface and subsequently were lowered to the growth temperature. Thus, the substrates were prepared for the film growth, of which procedures are specific to the respective films as described below:

For (Bi, Sb)$_2$Te$_3$, Cr$_x$(Bi$_{1-y}$Sb$_y$)$_{2-x}$Te$_3$ and V$_x$(Bi$_{1-y}$Sb$_y$)$_{2-x}$Te$_3$ growth, the growth temperature was about 160 to 200 °C. Bi, Sb and Te were co-evaporated, and Cr and V were selectively supplied. By monitoring beam fluxes, the total flux of Bi and Sb was maintained at 2.5 or 5.0 $\times 10^{-6}$ Pa and the ratio of Bi to Sb was carefully chosen for precise control of the chemical potential within the exchange gap. The Te was over-supplied (the Te/(Bi + Sb) ratio was close to 40) to suppress Te vacancies. Under this condition, the growth rate was about 0.1 to 0.2 nm min^{-1}. After the growth, annealing under exposure to Te was performed in situ at 380 °C for 30 min to make a smoother surface. For Cr$_2$Ge$_2$Te$_6$, Cr$_2$Si$_2$Te$_6$ growth, the growth

M. Mogi, *Quantized Phenomena of Transport and Magneto-Optics in Magnetic Topological Insulator Heterostructures*, Springer Theses, https://doi.org/10.1007/978-981-19-2137-7_2

Fig. 2.1 One of our MBE systems mainly used for the present thesis study

temperature was about 180 to 260 °C. Cr, Ge (Si) and Te were co-evaporated. The ratio of Ge (Si) to Cr was carefully chosen to realize the stoichiometric conditions. For more details, we discussed in Chap. 4. On taking out the films from the MBE chamber, a 3-nm-thick AlO_x capping layer was immediately deposited by atomic layer deposition (ALD) at room temperature to prevent deterioration of the films.

2.2   Device Fabrication

The films were patterned into Hall bars (Fig. 2.2a) and Corbino disks (Fig. 2.2b) by photolithography. In the photolithography process, the samples are spin-coated by hexamethyldisilazane (HMDS) to increase adhesion of photoresist and a solution of AZ1500 : anisole = 1 : 4 as photoresist, and then are exposed by ultraviolet (UV) light by using an UV maskless lithography system. We removed the photoresist exposed by the UV light by using a solution of AZ developer : H_2O = 1 : 1. Etching processes were performed by Ar ion-milling or by solutions of H_2O_2 : H_3PO_4 : H_2O = 1 : 1 : 8 and HCl : H_2O = 1 : 4. Electrical contact was made of Ti(5 nm)/Au(45 nm) by electron-beam evaporation. For gating the Hall bar device, we deposited AlO_x (~30 nm) by ALD and subsequently, Ti(5 nm)/Au(45 nm) electrodes were deposited by electron-beam evaporation.

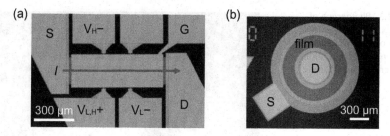

Fig. 2.2   a, b Optical microscope images of a Hall bar with a top-gate electrode (**a**) and a Corbino disk (**b**)

2.3   Electrical Transport and Magnetization Measurements

2.3.1   Electrical Transport Measurements

Electrical transport measurements were carried out in a Quantum Design PPMS with a ^{3}He probe in a variable temperature range of 0.5 to 300 K or with an adiabatic demagnetization refrigerator with a base temperature of 0.1 K, or in a dilution refrigerator with a base temperature of 40 mK. The electrical resistance was measured with the PPMS or with a standard lock-in technique. The excitation current was typically 10 to 100 nA to minimize the heating and the frequency was 3 Hz. The resistivities are derived from the measured resistances by $\rho_{xx} = R_{xx}/(L/W)$ and $\rho_{yx} = R_{yx}$, where W is the width of the sample, L is the longitudinal distance between voltage terminals, and R_{xx} and R_{yx} are the measured longitudinal resistance and Hall resistance, respectively. σ_{xx} and σ_{xy} were converted from ρ_{xx} and ρ_{yx} using tensor relations: $\sigma_{xx} = \rho_{xx}/(\rho_{xx}^2 + \rho_{yx}^2)$ and $\sigma_{xy} = \rho_{yx}/(\rho_{xx}^2 + \rho_{yx}^2)$ (Fig. 2.3).

2.3.2   Magnetization Measurements

Magnetization measurements were conducted in a superconducting quantum interference device (SQUID) magnetometer (MPMS). The diamagnetic contributions from the InP substrates were subtracted by their field-linear components above the saturation fields.

2.4   Polarized Neutron and X-Ray Reflectometry

Polarized neutron reflectometry (PNR) was performed at BL-17 SHARAKU of the Materials and Life Science Experimental Facility (MLF) in the Japan Proton Accelerator Research Complex (J-PARC), Tokai, Japan [1, 2]. The sample was loaded

Fig. 2.3 Photograph of
magnetic TI film devices
with gold wires contacted by
silver paste for a dilution
refrigerator measurement

into a closed cycle refrigerator and was cooled to 3 K. An electromagnet was used
to apply a magnetic field (1 T) along the in-plane direction of the film, which was
perpendicular to the incident neutron beam and neutron momentum transfer, Q_z. The
PNR spectra were measured by means of the time-of-flight technique with a pulsed
polychromatic incident neutron beam; the wavelength range was from 2.4 to 8.8 Å.
We selected three different incident angles (0.3°, 0.9° and 2.7°) to provide access to
the momentum transfer range from 0.08 to 2 nm^{-1}. A supermirror polarizer, guide-
field coils, and a spin flipper were employed to obtain the polarized incident neutron
beam, whose polarization direction was set to be parallel or antiparallel to the external magnetic field at the sample position. The beam polarization was approximately
98%. The intensities of reflected neutrons were measured without analyzing the spin
state of the neutrons and converted to the PNR spectra by the UTSUSEMI software
[3], in which effects of the imperfect beam polarization were corrected. The PNR
spectra were fitted using GenX software [4], assuming that the magnetic moments
of Cr were parallel to the magnetic field.

Complementary x-ray reflectometry (XRR) spectrum was measured by Bruker
D8 x-ray diffractometer at room temperature to determine layer thicknesses of the
sample. The sample used in the XRR measurement is identical to that used in the

PNR measurements. An incident x-ray beam with a wavelength of 1.5406 Å was obtained by Cu Kα radiation. The intensity of specular reflection was measured by varying the incident angle from 0.4° to 4.3° to cover Q_z range from 0.3 to 3 nm^{-1}.

2.5   Terahertz Magneto-Optical Spectroscopy

In the time-domain THz measurements, laser pulses with a wavelength of 800 nm and a duration of 100 fs generated by a mode-locked Ti: sapphire laser were split into two paths to emit and detect THz pulses by a photoconductive antenna and by a dipole antenna, respectively. The THz photon energy ($\hbar\omega$) range is 1 to 6 meV (0.2-1.5 THz in frequency) and is centered at approximately 2 meV. The samples were mounted on a copper stage with a hole of about 7 mm in diameter for the sufficient transmission of THz light and were cooled in a cryostat equipped with a superconducting magnet (7 T) and a ^{3}He probe insert down to 1 K. We measured transmitted THz electric fields $E_x(t)$ and $E_y(t)$ using two wire grid polarizers (parallel for the E_x measurements or orthogonal for the E_y measurements) across the samples (Fig. 2.4a). To eliminate the background signal except for the magneto-optical rotations, $E_y(t)$ was deduced from anti-symmetrizing waveforms of $E_y(+H)$ and $E_y(-H)$ ($E_y = (E_y(+H) - E_y(-H))/2$), where H is the magnetic field applied perpendicularly to the film planes. The energy spectra of the complex rotation angles are obtained by the Fourier analysis of $E_y(\omega)/E_x(\omega) = (\sin\theta(\omega) + i\eta(\omega)\cos\theta(\omega))/(\cos\theta(\omega) - i\eta(\omega)\sin\theta(\omega)) \sim \theta(\omega) + i\eta(\omega)$, where θ is the rotation angle and η is the ellipticity, for the small rotation angles. The magneto-optical Kerr measurement was performed with multiple reflections inside the substrate (InP). Of the waveform of the THz pulse transmitted through the sample (Fig. 2.4b), the second pulse comes from the multiple reflections inside the substrate after the first main pulse with a time delay $\Delta t = 2n_s d/c$, where d is the thickness of the InP substrate ($d = 360$ μm), c is the speed of light and n_s is the refractive index of the InP substrate. By using Δt, n_s is determined to be approximately 3.46 which

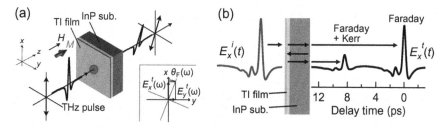

Fig. 2.4   a Schematic setup for the THz magneto-optical measurement for a TI thin film. **b** Typical waveform of the incident and transmitted THz pulse. The waveform of the transmitted electric field is divided into the first and second transmitted pulses with a time delay due to multiple reflections in the substrate

well agrees with the literature [5]. The second pulse experiences both the Faraday and Kerr effects when it is transmitted through the TI film and is reflected at the film-side surface of the substrate, respectively. Since the rotation angles of θ_F and θ_K are so small that the total rotation angle amounts to the sum of the Faraday and Kerr rotations. θ_K was obtained by subtracting the rotation angles for the first pulses from those of the second pulses. Under magnetic fields, we confirmed that the Kerr rotation does not come from the apparatus other than the sample and not from the InP substrate.

References

1. M. Takeda, D. Yamazaki, K. Soyama, Chin. J. Phys. **50**(2), 161 (2012)
2. K. Nakajima, Y. Kawakita, S. Itoh, J. Abe, K. Aizawa, H. Aoki, H. Endo, M. Fujita, K. Funakoshi, W. Gong et al., Quantum Beam Sci. **1**(3), 9 (2017)
3. Y. Inamura, T. Nakatani, J. Suzuki, T. Otomo, J. Phys. Soc. Jpn. **82**(Suppl. A), SA031 (2013)
4. M. Björck, G. Andersson, J. Appl. Crystallogr. **40**(6), 1174 (2007)
5. L. Alyabyeva, E. Zhukova, M. Belkin, B. Gorshunov, Sci. Rep. **7**(1), 7360 (2017)

Chapter 3
Magnetic Modulation Doping for Quantum Anomalous Hall Effect

3.1 Introduction

The QAH effect has attracted much attention owing to the generation of dissipation-less edge current without external magnetic fields. The first experimental discovery of the QAH effect was made by Chang et al. at an extremely low temperature of 30 mK. After that, the QAH effect has been reproducibly observed, but the observable temperature has been limited below 100 mK. Moreover, this temperature is two orders of magnitude lower than the Curie temperature T_C although the reported large magnetically induced gap seen by ARPES and STM is about 50 meV [1, 2]. Although the reasons for this discrepancy has been unclear yet, we speculate that those are (1) the inhomogeneity of the exchange coupling between the TI surface states and the doped magnetic moments and (2) the existence of the residual conduction at the bulk region formed by the structural and/or atomic inhomogeneity of the films. The problem (1) leads to the spatial fluctuation of the magnetic gap of the surface state as partly probed by STM [2]. This fluctuation diminishes the effective magnetic gap size; hence, the QAH state is adapted to be destroyed by thermal activation.

We have observed the QAH effect at 1 K (and possibly subsisting up to 2 K) by applying Cr modulation doping to TI $(Bi_{1-y}Sb_y)_2Te_3$ thin films. In the conventional modulation doping, a donor element, such as Si in GaAs/$Al_{1-x}Ga_xAs$ heterostructures, is introduced apart from the two-dimensional conduction channel at the heterointerface to reduce the ionized-impurity scattering, as contrasted by uniform doping [3, 4]. In the present magnetic modulation doping, on the other hand, high concentrated Cr doped thin layers (1 or 2 mono-layers) are introduced to the vicinity of the surface to make the spin-polarized surface state ferromagnetic, and to enhance the effective magnetic gap with suppression of the overall disorder.

M. Mogi, *Quantized Phenomena of Transport and Magneto-Optics in Magnetic Topological Insulator Heterostructures*, Springer Theses, https://doi.org/10.1007/978-981-19-2137-7_3

3.2 Higher-Temperature Quantum Anomalous Hall Effect in Modulation-Doped Topological Insulator

To explore the optimal modulation-doped structure of the TI films, we fabricated three devices as shown in Fig. 3.1a–c. The Bi/Sb ratio (y) was fixed at 0.78. Figure 3.1a shows the uniformly doped $Cr_x(Bi_{1-y}Sb_y)_{2-x}Te_3$ (CBST) ($x = 0.10$) film ("single-layer") showing the QAH effect at 50 mK. Figure 3.1b and c show the schematic structures to which magnetic modulation doping is applied; in the "tri-layer" structure (Fig. 3.1b), the top and bottom 1-nm surface layers are highly doped ($x = 0.46$), while in the "penta-layer" structure (Fig. 3.1c) Cr ($x = 0.46$) was doped similarly, but apart by 1 nm away from the topmost and bottommost surfaces. The T_C of the single-, tri-, and penta-layers are about 9, 25, and 25 K, respectively, which are estimated from the temperature dependence of R_{yx} under zero magnetic field (see Fig. 3.2d and its

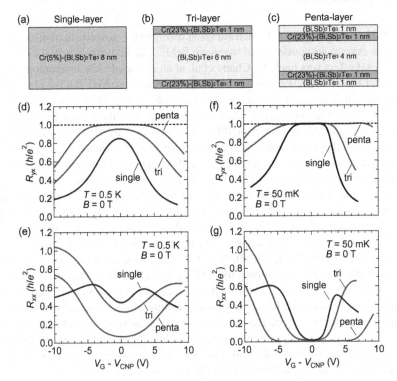

Fig. 3.1 a–c Schematic structures of the tested uniform and modulation magnetic doped films: **a** $Cr_x(Bi_{1-y}Sb_y)_{2-x}Te_3$ ($x = 0.10$, $y = 0.78$) single-layer, **b** ($x = 0.46$, $y = 0.78$) tri-layer, and **c** ($x = 0.46$, $y = 0.78$) penta-layer. **d–g** Gate voltage (V_G) dependence of **d, f** the Hall resistivity (R_{yx}) and **e, g** the longitudinal resistivity (R_{xx}) of the single-layer (gray line), the tri-layer (blue line), and penta-layer (red line) at 0.5 K and at 50 mK, respectively, at zero magnetic field ($B = 0$ T). Reprinted from [5], with the permission of AIP Publishing

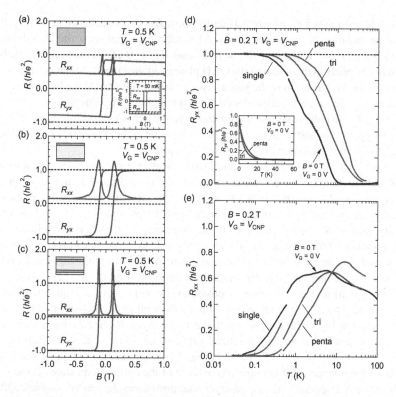

Fig. 3.2 a–c B dependence of the Hall (R_{yx}: blue line) and longitudinal (R_{xx}: red line) resistance in **a** the single-layer, **b** tri-layer, and **c** penta-layer at 0.5 K, tuned at the charge neutrality point by gating. The inset to **a** shows the QAH state of the single-layer at 50 mK. **d, e** T dependence of **d** R_{yx} and **e** R_{xx} in the respective structures under a magnetic field $B = 0.2$ T. In the single-layer, gate voltage (V_G) and B are not applied. The inset to **d** shows the behaviors for the tri-layer and penta-layer films without applying V_G and B (plotted on a linear scale). Reprinted from [5], with the permission of AIP Publishing

inset). We note that the nominal integrated amounts of Cr ions in the three films were tuned to almost the same values.

We display in Figs. 3.1d and e the gate voltage (V_G) dependence of the Hall (R_{yx}) and longitudinal (R_{xx}) resistance at 0.5 K at zero magnetic field ($B = 0$ T) after magnetic field cooling from $B = 2$ T. The charge neutrality points V_{CNP}, in which the tangent Hall angle (R_{yx}/R_{xx}) takes a maximum value, are 1.5 V for the single-layer, 0.9 V for the tri-layer, and 5.5 V for the penta-layer. In the single-layer (gray line), R_{yx} is about $0.85h/e^2$ and R_{xx} has a dip at $V_G = V_{CNP}$, indicating the incipient feature of the metallic chiral edge state in the QAH regime. The values of R_{yx} and R_{xx} at 0.5 K are comparable to the previous single-layer studies [6–11]. The tri-layer (blue line) exhibits the similar V_G dependence and approaches more closely to the QAH state than the single-layer case. In the penta-layer (red line), R_{yx} shows the quantized Hall plateau (h/e^2) at around $V_G = V_{CNP}$, indicating the appearance of the

QAH state even at 0.5 K. Figure 3.1f and g show the V_G dependence at 50 mK, the lowest temperature of the present study. All the three devices show the well-defined QAH state at $V_G = V_{CNP}$, i.e., $R_{yx} = h/e^2$ and $R_{xx} \sim 0 \ \Omega$. Besides, a remarkable difference appears in the width of the QAH plateau against the V_G variation; the wide plateau in the V_G variation in the penta-layer probably reflects (1) the large effective magnetic gap in the modulation-doped structure and (2) the difficulty of E_F tuning via gating. These two possibilities originate from the rich-Cr-doped layer providing a larger magnetic gap and parasitic states in the 1-nm layer, but the adverse implication is effectively minimized by the modulated structure.

To clarify the characteristics of the QAH effect in each structure, we displays the B and temperature T dependence under $V_G = V_{CNP}$. Figure 3.2a–c show the B dependence of R_{yx} and R_{xx} at 0.5 K in the single-layer, tri-layer, and penta-layer structures, respectively. Ferromagnetic hysteresis loops with the out-of-plane easy axis are seen in R_{yx} (blue line) for all the structures. At the coercive field ($\mu_0 H_c$) corresponding to the plateau-to-plateau transition, R_{xx} (red line) exhibits a peak, which probably suggests that the magnetic domain walls within the small multi-domains provide an objective effect to the edge current conduction [12]. The single-layer configuration (Fig. 3.2a) shows considerate deviations from the quantized value for R_{yx} or zero for R_{xx} at 0.5 K, while showing the well-defined QAH state at 50 mK (see the inset of Fig. 3.2a). In contrast to the single-layer, R_{yx} for the tri-layer for the penta-layer almost reaches the quantized value of $\pm h/e^2$ at 0.5 K. Note that the V_G dependence of R_{yx} at $B = 0$ T shown in Fig. 3.1d did not reach the quantized value for the tri-layer structure due to the coexistence of the anti-domain states in the zero magnetic field. Especially, the penta-layer structure shows the nearly complete QAH effect where $R_{yx} = \pm 1.00 h/e^2$ and $R_{yx} = 0.058 h/e^2$ even at 0.5 K. Figure 3.2d and e display the T dependence of R_{yx} and R_{xx} in the single-layer (gray), tri-layer (blue), and penta-layer (red) structures, respectively, where a low magnetic field (0.2 T) was applied so as to ensure the formation of a single magnetic domain state. As T decreases, ρ_{yx} increases below the T_C because the magnetization becomes large and the magnetic gap opens. In the order of the single, tri, and penta layers, T at which R_{yx} approaches the quantized value gets higher.

Another litmus test for the QAH state is the energy scale of the transport, which can be characterized by the activation energy determined from an Arrehenius plot of σ_{xx} as a function of $1/T$ (Fig. 3.3). $\log\sigma_{xx}$ linearly scales with $1/T$, indicating the thermal activated carrier transport. By fitting of them with a relation of $\sigma_{xx} \propto \exp(-\Delta/(2k_B T))$, Δ is notably getting higher in the order of the single, tri, and penta layers; Δ is 60 μeV for the single-layer, 90 μeV for the tri-layer, and 380 μeV for the penta-layer.

From these results, we conclude that the penta-layer is a suitable structure for observing higher-temperature QAH effect.

Toward the even higher-temperature QAH effect based on the penta-layer, we exemplify three penta-layer devices by varying the Cr concentration (x). For the x variation, Bi/Sb ratio (y) was also slightly tuned, so that V_{CNP} can be detected in V_G sweep range. The tested compositions in the penta-layer are $(x, y) = (0.46, 0.78)$, $(0.57, 0.74)$, and $(0.95, 0.68)$. The results for the $(x, y) = (0.46, 0.78)$ were already

Fig. 3.3 Arrhenius plots of
σ_{xx} at $B = 0.2$ T. Fitting by
$\sigma_{xx} \propto \exp(-\Delta/(2k_B T))$
gives the activation energy
$\Delta = 60, 90$, and $380 \, \mu$eV
for single-, tri-, and
penta-layers, respectively

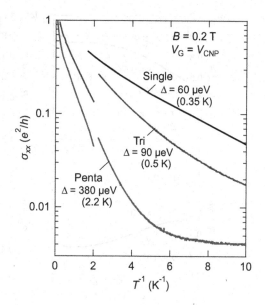

shown in Figs. 3.1 and 3.2. The respective T_C are about 25 K, 55 K, and 80 K that is judged from the T dependence of R_{yx}. Figure 3.4a and b show the V_G dependence of R_{yx} and R_{xx} at $T = 0.5$ K and $B = 0$ T. Among them, the $(x, y) = (0.57, 0.74)$ penta-layer presents a wide quantized plateau in Fig. 3.4a and a minimum value of R_{xx} in Fig. 3.4b. The highest Cr-doped $(x, y) = (0.95, 0.68)$ penta-layer shows worse behavior of R_{yx} and R_{xx} in the whole V_G region far from QAH state. Figure 3.4c and d show the B dependence of them at 0.5 K under $V_G = V_{CNP}$. It is to be noted that the higher Cr concentration provides a larger coercive field ($\mu_0 H_c$) with broad hysteresis. Thus, there is an optimum Cr concentration in the penta-layer in this present study, $(x, y) = (0.57, 0.74)$, presenting a well-defined QAH effect at 0.5 K, where the residual R_{xx} at zero magnetic field is as small as $0.017 \, h/e^2$.

In Fig. 3.5, the B dependence of R_{yx} and R_{xx} for the optimized penta-layer $(x, y) = (0.57, 0.74)$ at $V_G = V_{CNP}$ is compared by increasing T: 0.5 K, 1 K, 2 K, and 4.2 K. Quantized R_{yx} is observed up to 1 K, where the residual R_{xx} is $0.081 h/e^2$, which is a little higher than that at 0.5 K ($0.017 \, h/e^2$). At 2 K, R_{yx} is still close to the quantized value, $\pm 0.97 h/e^2$. Then, at 4.2 K, it goes away from QAH state ($R_{yx} = \pm 0.87 h/e^2$), but the large Hall angle still remains. To observe the QAH effect beyond the liquid ^{4}He temperature, fine tuning of Cr concentration (x) and Bi/Sb ratio (y) to suppress disorder and/or doping another magnetic element, such as V which can raise the Curie temperature as compared with the similar concentration of Cr [10, 13], is required in the future research.

We finally discuss the origin for the improvement of the observable temperature of the QAH effect in the magnetic modulation doped structures. Since the Hall conductivity in QAH regime is observed with the definite quantized value of $\pm e^2/h$, i.e., twice a half quantized value, it is likely that the modulation-doping at the vicinity

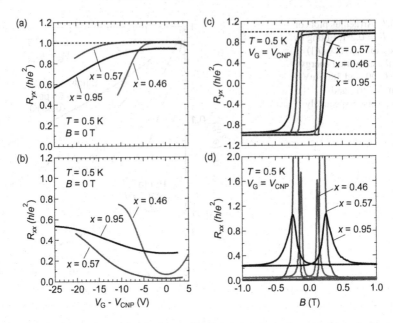

Fig. 3.4 **a, b** V_G dependence of **a** the R_{yx} and **b** R_{xx} resistance in the penta-layer devices at 0.5 K in zero magnetic field ($B = 0$ T). The results for three devices with the different Cr and Bi/Sb compositions in the penta-layer structures; $(x, y) = (0.46, 0.78)$ (red line), $(0.57, 0.74)$ (blue line), and $(0.95, 0.68)$ (gray line). **c, d** B dependence of **c** R_{yx} and **d** R_{xx} in the respective compositions at 0.5 K, tuned at the charge neutrality point by gating to $V_G = V_{CNP}$. Reprinted from [5], with the permission of AIP Publishing

of the top and bottom surfaces keeps the nature of the three-dimensional topological insulator for the whole film. In the ideal QAH state, there is no bulk conducting state in the surface magnetic gap $\Delta \propto J M_z$, where J is the exchange coupling constant between the surface electrons and magnetic dopant ions and M_z is the magnetic moments perpendicular to the surface [14, 15]. In real electrical measurements, however, the size of the magnetic gap is effectively reduced by the inhomogeneity of the Cr concentration and the bulk conducting states. Since the M_z is proportional to Cr concentration, the effective magnetic gap should be subject to the spatial fluctuation in the magnetic doped layer, which diminishes the observable temperature of QAH effect far below the T_C. Magnetic modulation doping makes it possible to dope high concentration of Cr and provide strong magnetic coupling between the surface state and the near-by magnetic moments suppressing bulk conduction, which leads to acquisition of the large effective magnetic gap. This is why the observable temperature is higher in tri-layer and penta-layer than the single-layer. By contrast, the J in the penta-layer would be smaller than that in the trilayer, taking the doped position of magnetic moments into account [16]. In such Cr-doped (Bi, Sb)$_2$Te$_3$ films, the ferromagnetic interaction itself is primarily mediated not by the surface electrons (RKKY interaction) but by the bulk property, i.e., the van Vleck mechanism [17].

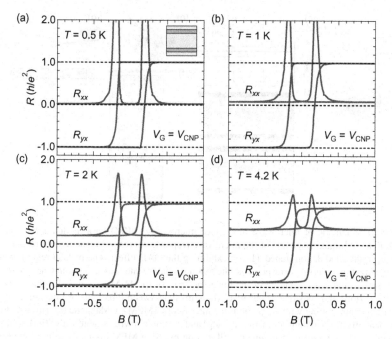

Fig. 3.5 **a–d** B dependence of R_{yx} and R_{xx} in the penta-layer device ($x = 0.57$, $y = 0.74$) at $T = 0.5$ K (**a**), 1 K (**b**), 2 K (**c**) and 4.2 K (**d**). V_G was tuned at the V_{CNP}. Reprinted from [5], with the permission of AIP Publishing

Therefore, the difference in the magnetic gap would originate from the difference in the J not M_z. However, the QAH state in the penta-layer structure is more stabilized than that in the tri-layer, as shown in Figs. 3.1 and 3.2, while their T_C are comparable. Thus, we can conclude that the important factor to enhance the thermal stability of the QAH state is not only the size of the effective magnetic gap but also its spatial homogeneity, which is improved by the remote doping of the magnetic moments from the surface layer in the penta-layer structure.

3.3   Detection of Zero-Field Chiral Edge Conduction

The higher temperature realization of the QAH effect enables us to probe it with various experimental techniques such as scanning probe microscopy. One of the examples is the imaging of QAH chiral edge conduction by using microwave impedance microscopy (MIM) [18]. One other example is the detection of chiral edge conduction on magnetic domain walls in the QAH state by using magnetic force microscopy [19].

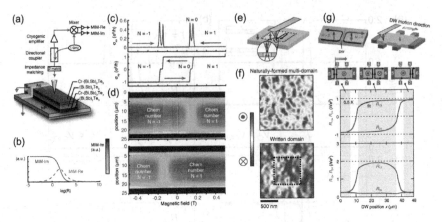

Fig. 3.6 a Schematic of microwave impedance microscopy (MIM) on a modulation-doped TI film. **b** Real and imaginary parts of the microwave response curves. **c** Magnetic field response of transport in the modulation-doped TI film exhibiting the QAH effect. The plateau region around the magnetization reversal corresponds to the axion insulator state as discussed in Chap. 5. **d** The corresponding real space imaging of the imaginary part of the MIM responses for the 'negative to positive' (top) and 'positive to negative' (bottom) field-sweep directions. **e** Illustration of the magnetic domain writing with magnetic force microscopy (MFM). **f** Magnetic domain structures for the naturally-formed multi-domain state (top) and the domain written state with MFM. **g** In-situ transport experiment during the domain wall motion by using MFM domain writing, suggesting the existence of the quantized chiral edge conduction on the domain wall. Panels **a–d** are reprinted with permission from [18], PNAS. Panels **e–g** are reprinted from [19] with permission from AAAS

For the MIM measurement, the complex reflectance of microwaves (at a frequency of 1 GHz) sent to the tip is measured (Fig. 3.6a). The responses provide information on the conductivity and the capacitance of the sample (Fig. 3.6b) at a local region (the typical spatial resolution ~ 100 nm) so that this MIM is a powerful technique to visualize the chiral edge conduction. The measurements were performed at 450 mK, exhibiting the QAH effect almost without any dissipation (or zero σ_{xx}) as shown in Fig. 3.6c. Figure 3.6d presents the real space imaging of the QAH state under the variation of magnetic fields. The imaginary part of the MIM response, which is sensitive to the conductivity of the sample (Fig. 3.6b), displays a noticeable enhancement of conductivity only at the edge, confirming the chiral edge states localized at the edge ($\sim \mu$m) in the QAH state.

MFM is usually used for measuring magnetic domain structures for magnetic materials by utilizing the magnetic (e.g. Co-coated) tip. In this experiment, the magnetic tip is used for writing the magnetic domains by using the stray fields from the tip (Fig. 3.6e). This technique demonstrates the domain writing and the imaging for the magnetic TI films as shown in Fig. 3.6f. On the domain walls of magnetic TIs, we expect the emergence of the two chiral edge channels propagating in parallel because the domain wall corresponds to the boundary of a discontinuous change of Chern numbers between $C = 1$ and -1. Hence, by creating a domain wall on a

QAH state, it affects the electrical transport. Let us consider the situation that one domain wall lies perpendicular to the current flow and between the voltage terminals as schematically shown in Fig. 3.6g (center). At the domain wall, two channels of the chiral edge states meet and travel in the same direction. They intermix and equilibrate with each other so that the potentials downstream along the domain wall become equal; hence R_{34} becomes zero. By contrast, R_{12} becomes $2h/e^2$ because the resistance between the contacts 1 to 3 and that between the contacts 2 to 4 take the opposite signs of the quantized value ($+h/e^2$ and $-h/e^2$, respectively). Taking $R_{34} = 0$ into consideration, the value of $R_{12} = 2h/e^2$ is explained. These resistance values can also be simulated by using Landauer-Büttiker formalism [20]. The above expectation was actually observed in the experiment shown in Fig. 3.6g (bottom), evidencing the existence of the chiral edge states on the domain wall.

3.3.1 Other QAH Materials

As a final remark, very recently, two new QAH systems have appeared, of which the observable temperature is several kelvins, comparable to the present modulation-doped TIs. One is an intrinsic antiferromagnetic TI of $MnBi_2Te_4$ and the other is 'non-magnetic' twisted bilayer graphene.

In $MnBi_2Te_4$ with a layered crystal structure similar to Bi_2Te_3, Mn ions are already incorporated in the composition and the system remains topologically nontrivial. The magnetic order of $MnBi_2Te_4$ is a layered antiferromagnetic state with perpendicular anisotropy [21, 22]. Because of the antiferromagnetic order, spontaneous anomalous Hall effect cannot be expected in a bulk crystal form. However, by exfoliating the crystal down to several layers, odd or even number of layers are switched to the QAH phase or a zero anomalous Hall state [21, 23]. Although the QAH effect has not been observed at zero magnetic field possibly due to still low bulk resistivity, a magnetic-field-assisted QAH effect was observed [24–26].

In twisted-bilayer graphene, one of the two graphene sheets in a bilayer graphene is rotated with respect to the other sheet at around 1°, leading to the emergence of extremely flat moiré superlattice minibands [27]. When E_F is tuned at the half-filling of the moiré flat bands, unusual insulating states due to strong electron-electron interactions have been found [28]. Strikingly, superconductivity with T_C of several kelvins shows up when carriers are slightly doped from the half-filling [29]. The QAH phase appears at 3/4 filling state without magnetic atoms [30, 31], where an additional symmetry breaking from the adjacent hBN layer drives the valley polarization, leading to the emergence of orbital magnetization breaking time-reversal symmetry. The quantized Hall resistance at zero magnetic field with hysteresis behaviors are clearly observed. The quantization persists up to 4 K. Furthermore, perhaps owing to the cleanliness of the graphene-based superlattices, the stability of the QAH effect (about 30 K as judged from the thermal activation energy) exceeds the ordering temperature of about 7 K in contrast to the magnetically doped TI systems.

3.4　Conclusion

To conclude, we have investigated the effects of magnetic modulation doping on TIs. The QAH effect was successfully stabilized up to 1K by inserting a 1 nm thick Cr doped layer near the top and bottom surfaces of the 8-nm-thick BST films. This temperature is more than an order of magnitude higher than that for the uniformly Cr-doped BST single-layer film. The effect of the modulation doping is that the magnetic gap increases with increasing Cr doping concentration and the doping-induced disorder in the conducting surface states decreases. We believe that this method based on TI superlattice engineering opens the way to study of the zero-field dissipationless currents and for the exploration of exotic topological phenomena.

References

1. Y. Chen, J.H. Chu, J. Analytis, Z. Liu, K. Igarashi, H.H. Kuo, X. Qi, S.K. Mo, R. Moore, D. Lu et al., Science **329**(5992), 659 (2010)
2. I. Lee, C.K. Kim, J. Lee, S.J. Billinge, R. Zhong, J.A. Schneeloch, T. Liu, T. Valla, J.M. Tranquada, G. Gu et al., Proc. Natl. Acad. Sci. U.S.A. **112**(5), 1316 (2015)
3. R. Dingle, H. Störmer, A. Gossard, W. Wiegmann, Appl. Phys. Lett. **33**(7), 665 (1978)
4. H. Störmer, A. Pinczuk, A. Gossard, W. Wiegmann, Appl. Phys. Lett. **38**(9), 691 (1981)
5. M. Mogi, R. Yoshimi, A. Tsukazaki, K. Yasuda, Y. Kozuka, K. Takahashi, M. Kawasaki, Y. Tokura, Appl. Phys. Lett. **107**(18), 182401 (2015)
6. C.Z. Chang, J. Zhang, X. Feng, J. Shen, Z. Zhang, M. Guo, K. Li, Y. Ou, P. Wei, L.L. Wang et al., Science **340**(6129), 167 (2013)
7. J. Checkelsky, R. Yoshimi, A. Tsukazaki, K. Takahashi, Y. Kozuka, J. Falson, M. Kawasaki, Y. Tokura, Nat. Phys. **10**, 731 (2014). https://doi.org/10.1038/nphys3053
8. X. Kou, S.T. Guo, Y. Fan, L. Pan, M. Lang, Y. Jiang, Q. Shao, T. Nie, K. Murata, J. Tang, Y. Wang, L. He, T.K. Lee, W.L. Lee, K.L. Wang, Phys. Rev. Lett. **113**, 137201 (2014). https://doi.org/10.1103/PhysRevLett.113.137201. https://link.aps.org/doi/10.1103/PhysRevLett.113.137201
9. A. Kandala, A. Richardella, S. Kempinger, C.X. Liu, N. Samarth, Nat. Commun. **6**, 7434 (2015). https://doi.org/10.1038/ncomms8434
10. C.Z. Chang, W. Zhao, D.Y. Kim, H. Zhang, B.A. Assaf, D. Heiman, S.C. Zhang, C. Liu, M.H. Chan, J.S. Moodera, Nat. Mater. **14**(5), 473 (2015)
11. S. Grauer, S. Schreyeck, M. Winnerlein, K. Brunner, C. Gould, L. Molenkamp, Phys. Rev. B **92**(20), 201304 (2015)
12. J. Wang, B. Lian, S.C. Zhang, Phys. Rev. B **89**, 085106 (2014). https://doi.org/10.1103/PhysRevB.89.085106. https://link.aps.org/doi/10.1103/PhysRevB.89.085106
13. Y. Ou, C. Liu, G. Jiang, Y. Feng, D. Zhao, W. Wu, X.X. Wang, W. Li, C. Song, L.L. Wang et al., Adv. Mater. **30**(1), 1703062 (2018)
14. Q. Liu, C.X. Liu, C. Xu, X.L. Qi, S.C. Zhang, Phys. Rev. Lett. **102**(15), 156603 (2009)
15. K. Nomura, N. Nagaosa, Phys. Rev. Lett. **106**, 166802 (2011). https://doi.org/10.1103/PhysRevLett.106.166802. https://link.aps.org/doi/10.1103/PhysRevLett.106.166802
16. F. Zhang, C.L. Kane, E.J. Mele, Phys. Rev. B **86**, 081303 (2012). https://doi.org/10.1103/PhysRevB.86.081303. https://link.aps.org/doi/10.1103/PhysRevB.86.081303
17. R. Yu, W. Zhang, H.J. Zhang, S.C. Zhang, X. Dai, Z. Fang, Science **329**(5987), 61 (2010)
18. M. Allen, Y. Cui, E.Y. Ma, M. Mogi, M. Kawamura, I.C. Fulga, D. Goldhaber-Gordon, Y. Tokura, Z.X. Shen, Proc. Natl. Acad. Sci. U.S.A. **116**(29), 14511 (2019)
19. K. Yasuda, M. Mogi, R. Yoshimi, A. Tsukazaki, K. Takahashi, M. Kawasaki, F. Kagawa, Y. Tokura, Science **358**(6368), 1311 (2017)

20. M. Büttiker, Phys. Rev. B **38**, 9375, 081303 (1988). https://doi.org/10.1103/PhysRevB.38. 9375. https://link.aps.org/doi/10.1103/PhysRevB.38.9375

21. J. Li, Y. Li, S. Du, Z. Wang, B.L. Gu, S.C. Zhang, K. He, W. Duan, Y. Xu, Sci. Adv. **5**(6), eaaw5685 (2019)

22. M. Otrokov, I. Klimovskikh, H. Bentmann, D. Estyunin, A. Zeugner, Z. Aliev, S. Gaß, A. Wolter, A. Koroleva, A. Shikin et al., Nature **576**(7787), 416 (2019)

23. M. Otrokov, I. Rusinov, M. Blanco-Rey, M. Hoffmann, A.Y. Vyazovskaya, S. Eremeev, A. Ernst, P. Echenique, A. Arnau, E. Chulkov, Phys. Rev. Lett. **122**(10), 107202 (2019)

24. Y. Gong, J. Guo, J. Li, K. Zhu, M. Liao, X. Liu, Q. Zhang, L. Gu, L. Tang, X. Feng et al., Chin. Phys. Lett. **36**(7), 076801 (2019)

25. Y. Deng, Y. Yu, M.Z. Shi, J. Wang, X.H. Chen, Y. Zhang (2019). arXiv preprint arXiv:1904.11468

26. C. Liu, Y. Wang, H. Li, Y. Wu, Y. Li, J. Li, K. He, Y. Xu, J. Zhang, Y. Wang (2019). arXiv preprint arXiv:1905.00715

27. R. Bistritzer, A.H. MacDonald, Proc. Natl. Acad. Sci. USA **108**(30), 12233, 081303 (2011). https://doi.org/10.1073/pnas.1108174108. https://www.pnas.org/content/108/30/12233

28. Y. Cao, V. Fatemi, A. Demir, S. Fang, S.L. Tomarken, J.Y. Luo, J.D. Sanchez-Yamagishi, K. Watanabe, T. Taniguchi, E. Kaxiras et al., Nature **556**(7699), 80 (2018)

29. Y. Cao, V. Fatemi, S. Fang, K. Watanabe, T. Taniguchi, E. Kaxiras, P. Jarillo-Herrero, Nature **556**(7699), 43 (2018)

30. A. Sharpe, E. Fox, A. Barnard, J. Finney, K. Watanabe, T. Taniguchi, M. Kastner, D. Goldhaber-Gordon, Science **365**, eaaw3780 (2019). https://doi.org/10.1126/science.aaw3780

31. M. Serlin, C. Tschirhart, H. Polshyn, Y. Zhang, J. Zhu, K. Watanabe, T. Taniguchi, L. Balents, A. Young, Science (2019)

Chapter 4
Magnetic Proximity Induced Quantum Anomalous Hall Effect

4.1 Introduction

Besides magnetic doping, stacking a ferromagnet on TIs is anticipated to be an appealing approach to induce the exchange gap for the topological surface state. Such a magnetic proximity effect may be a way to produce the spatially uniform exchange interaction between TI surface states and magnetic moments, which can lead to enhance the effective size of the exchange gap. To realize the efficient exchange coupling, the material choice for ferromagnetic insulators (FMIs) is a important problem to induce the efficient exchange coupling with less disorder; candidates for the FMIs stacking on the TI are of a great variety. Indeed, various FMI/TI heterostructures have been proposed theoretically and synthesized to date [1–15]. Whereas these studies have demonstrated some potential magnetoelectronic responses, the anomalous Hall effect [3–10], and unconventional surface magnetization [11, 12], even at room temperature, the magnitude of the response and the coupling strength to the TI surface state has been far smaller than expected.

In this chapter, we chose $Cr_2Ge_2Te_6$ and $Cr_2Si_2Te_6$ as FMIs to induce efficient magnetic proximity effect. By fabricating heterostructures composed of $(Bi, Sb)_2Te_3$ and $Cr_2Ge_2Te_6$ or $Cr_2Si_2Te_6$, we study the anomalous Hall effect and the related spintronic properties including current-induced magnetization switching by utilizing the spin-momentum locking of the TI surface states and its coupling with FMIs.

4.2 Growth of van der Waals Ferromagnetic Insulators

$Cr_2Ge_2Te_6$ (CGT), which is one of the insulating ferromagnets, is a promising vdW material compatible with such layered chalcogenides. Recently, such magnetic vdW materials have gathered renewed interest since the long-range ferromagnetic ordering survives even down to a two-dimensional monolayer or a few monolayers in magnetic insulators, such as $Cr_2Ge_2Te_6$[16] and CrI_3 [17]. The characteristic magnetism leads to new device concepts based on the vdW structure. By applying an exfoliation technique, the electric field effect [18–21], giant mangetoresistance [22, 23], and optical control [24] in vdW crystal-based heterostructures have been rapidly explored, which envisions high-performance spintronic functionalities in two-dimensional materials [25, 26]. Significantly, concomitant ferromagnetism and insulating features in the magnetic vdW materials provide a feasible arena for exploration of the proximity effect in the stacked heterostructures with other vdW materials such as graphene, transition metal dichalcogenides, and topological insulators. For instance, the vdW magnetic insulator can magnetize the surface of the topological insulator by the proximity effect, potentially driving them into the gap formation of the surface Dirac bands and/or the efficient magnetization control via spin-momentum locking [27]. Thus, CGT, which is one of the insulating ferromagnets, is a promising vdW material compatible with layered chalcogenides based topological insulators [7, 28]. However, MBE-growth of CGT has remained challenging toward the formation of vdW-based heterostructure.

In this section, we explore the MBE thin-film growth of the ferromagnetic insulator CGT. In the systematic investigation with various molecular-beam flux ratios of Ge to Cr, we fabricated a series of mixed crystals from the metallic Cr_2Te_3 [29–31] to the insulating CGT in the Cr-Ge-Te ternary system. The fine tuning of the Ge:Cr flux ratio is critical to obtain the high resistivity and the bulk-comparable Curie temperature (T_C) in the ferromagnetic CGT films. Interestingly, the nearly stoichiometric CGT thin film has a large spontaneous magnetization with out-of-plane easy axis.

The crystal structure of CGT is rhombohedral with vdW stacking of hexagonal layers, as shown in Fig. 4.1a [32]. Such a layered crystal structure is expected to be suitable to form an abrupt interface with other vdW materials. Here we applied the BST (Fig. 4.1c) as a buffer layer to form rhombohedral Te-based layer stacking. Even though the BST has a conductive surface when it is thick enough (> 6 nm), the ultrathin layer ($1 \sim 2$ nm) becomes insulating due to the hybridization between the top and bottom surface states [33]. In contrast, Cr_2Te_3 (Fig. 4.1b) has a trigonal structure which is not a vdW layered compound but contains a similar Te atomic arrangement in the ab-plane. The in-plane orientation of CGT (or Cr_2Te_3) on BST is expected $30°$ with respect to the unit cell of BST to match the Te arrangement in the ab-plane, as shown in Fig. 4.1d and e.

We grew Cr-Ge-Te alloy thin films on the 1-2 nm BST buffer layers on InP(111) substrates, as shown in Fig. 4.1f. Structural characterizations for 36-nm-thick films controlled with the Ge;Cr flux ratio ($P_{Ge}/P_{Cr} = 0, 1, 2, 3, 4$) were performed by XRD as shown in Fig. 4.1g. For the film grown under $P_{Ge}/P_{Cr} = 0$ (bottom light blue line), we observed the diffraction peaks at (002), (004), and (008), which indicates c-axis oriented Cr_2Te_3 with a trigonal structure. For the growth condition of $P_{Ge}/P_{Cr} = 4$ (top red line), the film exhibits diffraction peaks at (003), (006), and (0012), corresponding to the rhombohedral CGT. In between these two films, the diffraction peak around $14°$ systematically shifts to the smaller angle with increasing the flux ratio of P_{Ge}/P_{Cr}, indicating the elongation of the lattice spacing along the c-axis direction with the variation of the composition (Fig. 4.1h). If the alloyed

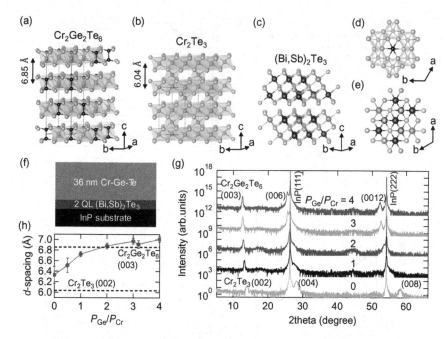

Fig. 4.1 **a–c** Illustrations of crystal structures for CGT (**a**), Cr_2Te_3 (**b**), and (Bi, Sb)$_2$Te$_3$ (**c**). **d, e** In-plane lattice configurations for CGT (**d**) and BST (**e**). When CGT is grown on the BST buffer layer, the unit cells of them are rotated by 30° in-plane. **f** Cross-sectional illustration of the Cr-Ge-Te ternary thin film on the BST/InP substrate. **g** XRD patterns of 36-nm-thick films with various Ge: Cr flux ratios ($P_{Ge}/P_{Cr} = 0, 1, 2, 3, 4$). Broad peaks around 17° and 44° possibly come from the diffraction at (006) and (0015) of the BST buffer layer. **h** d-spacing obtained from **g** as a function of P_{Ge}/P_{Cr}. The upper and lower broken lines represent the bulk values for CGT and Cr_2Te_3, respectively. Reprinted from [34] by The Author(s) licensed under CC BY 4.0

crystals are in the same crystal structure, such a systematic peak shift would be understood as Vegards's law. However, the end compounds have different crystal structures in the present case. For the origin of the gradual shift in the d-spacing, we speculate that the crystal structure gradually modifies from Fig. 4.1b to a. In other words, the gradual change from Cr_2Te_3 to $Cr_2Ge_2Te_6$ can be viewed as Ge dimer incorporation in each monolayer with the formation of the vdW gap, accompanied by the reduction of Cr in the Te-Te interlayer region.

We then studied the magnetization and transport properties. Figure 4.2a shows the temperature dependent magnetization curves, where the magnetic fields $B_\perp(\|c) = 50$ mT are applied perpendicular to the films. While all the films exhibit ferromagnetic behaviors, T_C widely changes from 170 to 80 K with increasing P_{Ge}/P_{Cr}. The relatively high T_C's (~170 K) for the films with $P_{Ge}/P_{Cr} \sim 0$ and 1 are close to the bulk and thin-film values of Cr_2Te_3 [29–31], while the low T_C's (~ 80 K) for the films with $P_{Ge}/P_{Cr} \sim 3$ and 4 are comparable to that of the bulk value (60–70 K) of CGT [16, 18, 28, 32, 35]. The variation of the magnetic moments at lowest temperature

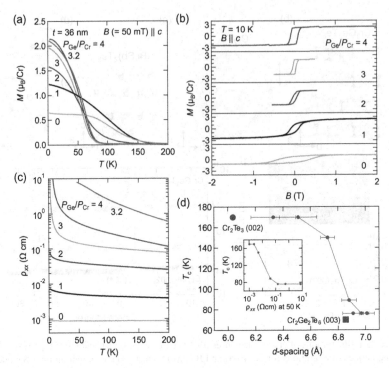

Fig. 4.2 a Temperature (T) dependence of the magnetization (M) for the 36-nm-thick films grown with various P_{Ge}/P_{Cr} ratios from 0 to 4 under a magnetic field of $B = 50$ mT applied along the c-axis. **b** Magnetization curves taken at a temperature at $T = 10$ K for the respective films shown in **a**. **c** T dependence of resistivity (ρ_{xx}) for 36-nm-thick films grown with $P_{Ge}/P_{Cr} = 0, 1, 2, 3, 3.2, 4$. **d** Curie temperature (T_C) as a function of the d-spacing as obtained from the XRD patterns shown in Fig. 4.1d as compared with bulk values of Cr_2Te_3 [29] (blue circle) and CGT [32] (blue square). The inset shows T_C as a function of ρ_{xx} at $T = 50$ K obtained from **c**. Reprinted from [34] by The Author(s) licensed under CC BY 4.0

(2 K) among these films comes from the differences in spontaneous magnetization at zero field as well as the perpendicular magnetic anisotropy. We note that the slightly higher T_C of 80 K for the films may be related to the enhancement of perpendicular magnetic anisotropy would result in the suppression of thermal fluctuations for low dimensional magnetism in the MBE-grown films. Magnetization curves at $T = 10$ K are plotted in Fig. 4.2b. We observed the well-developed hysteresis loops in all the films, indicating the long-range ferromagnetic order with perpendicular magnetic anisotropy. Moreover, the rectangular shape of the hysteresis loop becomes clearer with increasing P_{Ge}/P_{Cr}, namely, toward the CGT end. The saturation magnetization (M_s) for the four films except for $P_{Ge}/P_{Cr} = 0$ (Cr_2Te_3 film) are around 2.0 to 2.4 μ_B /Cr atom, being consistent with the bulk value of 2.2 to 3 μ_B /Cr atom in the CGT crystals [28, 32, 35]. Additionally, the reduction of M_s to about $1 - 2\mu_B$ /Cr

for the Cr_2Te_3 film can be explained by a small antiferromagnetic contribution from Cr at the Te-Te interlayer region [29, 30].

Figure 4.2c shows the temperature (T) dependence of longitudinal resistivity (ρ_{xx}), which strongly depends on the flux ratio of Ge to Cr. In the film grown under $P_{Ge}/P_{Cr} = 0$, less T dependence of ρ_{xx} is observed in accord with the reported features for Cr_2Te_3 bulk crystals and thin films [29–31]. With increasing Ge flux, on the other hand, ρ_{xx} increases at a whole T range. Indeed, the insulating behavior is pronounced when P_{Ge}/P_{Cr} is optimized to be 3.2. Judging from M_s of 2.4 μ_B/Cr, T_C of about 80 K and the insulating behavior, the quality of the phase-purity of Cr-Ge-Te films grown under $P_{Ge}/P_{Cr} = 3 - 4$ is comparable to that of the bulk single crystal of CGT. In Fig. 4.2d, T_C for the Cr-Ge-Te films is plotted as a function of the d-spacing along the c-axis direction, comparing with bulk values for Cr_2Te_3 (blue circle) [29] and CGT (blue square) [32]. T_C of films with $P_{Ge}/P_{Cr} = 0$ and $P_{Ge}/P_{Cr} = 3 - 4$ correspond to that of Cr_2Te_3 and CGT, respectively. However, the T_C values for the films grown under various P_{Ge}/P_{Cr} could not be interpreted with a linear relationship against the d-spacing between the two bulk values for Cr_2Te_3 and CGT. The observed abrupt change may originate not only from the structural change in which Cr atoms are replaced by Ge dimers to form the vdW gap with the weakened Cr-Cr magnetic interaction but also from the carrier-mediated ferromagnetic interaction occurring in metallic Cr_2Te_3. To see the possible latter effect, we plot T_C versus resistivity $(\rho_{xx}$ at 50 K) for the respective films in the inset of Fig. 4.2d, confirming the strong correlation between the two quantities.

Hereafter, we focus on the structure and magnetic properties of the highly resistive Cr-Ge-Te film grown under $P_{Ge}/P_{Cr} = 3.2$. Figure 4.3a, b, and c show the high-angle annular dark-field (HAADF) images taken by a scanning transmission electron microscopy (STEM). Te, In, Bi, and Sb are mainly observed due to their large atomic numbers compared with Cr, Ge, and P. In the fairly large area shown in Fig. 4.3a and b, there is discerned a well-ordered layer-stacking structure with neither dislocation nor segregation. In the magnified image (Fig. 4.3c), a contrast between the interlayer and the intralayer of Cr-Ge-Te layers can be observed, suggesting that the interlayer correspond to the vdW gap. Furthermore, a sharp interface is observed between Cr-Ge-Te and $(Bi, Sb)_2Te_3$ layers with the expected crystal-orientation relationship. By performing the Fourier transformation along the lateral direction of the STEM image (Fig. 4.3c) the lateral atomic distance of each layer was estimated as shown in Fig. 4.3d). The observed in-plane Te-Te distances of Cr-Ge-Te [= 6.89(5) Å] and BST [= 7.38(5) Å] agree well with the reported bulk values of CGT and BST with the identical Sb composition [28, 32, 36]. At the interface, the Fourier-transformed color plot shows a sharp change in the lateral atomic distance, reflecting the character of the vdW interface. Figure 4.3e–i show uniform distribution of the Cr, Ge, Te, In, and P atoms as probed by STEM energy dispersive x-ray spectroscopy (EDX). In Fig. 4.3(j), the averaged composition fraction profiles of Ge and Te against Cr along the growth direction are plotted. The averaged fractions of Ge/Cr and Te/Cr are approximately unity and three, respectively. Judging from the structural and compositional characterizations, the Cr-Ge-Te film grown under $P_{Ge}/P_{Cr} = 3.2$ can be identified to be the CGT phase.

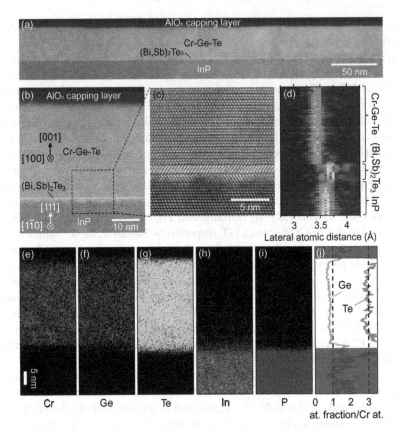

Fig. 4.3 a, b Cross-sectional HAADF-STEM images of a 36-nm-thick Cr-Ge-Te film on the BST/InP substrate, which is protected by an AlO$_x$ capping layer. The scale bars indicate 50 nm **a** and 10 nm **b** in length. **c** Expanded image of the Cr-Ge-Te/BST interface. The scale bar indicates to 5 nm. **d** Lateral atomic distance of each layer obtained by Fourier transformation of **c** plotted along the growth direction. **e–i** Elemental distributions of Cr (**e**), Ge (**f**), Te (**g**), In (**h**), P (**i**) obtained by EDX. The scale bar indicates to 5 nm. **j** Averaged compositional fraction profiles for Ge and Te atoms normalized by Cr atoms along the growth direction deduced from **e–g**. Gray-shaded regions correspond to the InP substrate (lower) and the AlO$_x$ capping layer (upper). Reprinted from [34] by The Author(s) licensed under CC BY 4.0

Finally, we investigated the thickness (t) dependence of magnetic properties for the CGT films grown under the nearly stoichiometric condition with $P_{Ge}/P_{Cr} = 3.2$ (Fig. 4.4a); the film thickness ($t = 6, 18, 36$ nm) was determined with the analysis of the x-ray reflectivity fringes as shown in Fig. 4.4b. Figure 4.4c displays the T dependence of the magnetization for the films by applying $B = 50$ mT compared with the bulk CGT single crystal. The T_C values of the films are almost constant about 80 K, as shown in the inset of Fig. 4.4c. Only for the $t = 6$ nm film, the T_C slightly decreases possibly due to a dimensionality effect as also seen in the mechanically exfoliated thin flakes. An important finding is the large spontaneous

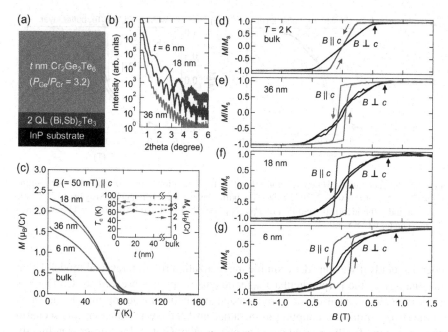

Fig. 4.4 a Cross-sectional schematic of t-nm-thick $Cr_2Ge_2Te_6$ grown with a flux ratio of $P_{Ge}/P_{Cr} = 3.2$ on the (Bi, Sb)$_2$Te$_3$/InP substrate. **b** X-ray reflectivity scans and deduced sample thicknesses. **c** Temperature (T) dependence of magnetization (M). Inset: thickness dependence of the Curie temperature (T_C) (shown by red circles) and saturation magnetization (M_s) (shown by blue circles). Magnetic field (B) dependence of M normalized by M_s of bulk crystal (**d**), 36 nm (**e**), 18 nm (**f**), and 6 nm (**g**) films measured at $T = 2$ K. Magnetic field is applied along the c-axis (shown in corresponding color) and perpendicular to the c-axis (shown in black). Black arrows for $B \perp c$ indicate the saturation field. Irregular behavior in the $B\|c$ magnetization curve in **g** around ±0.1 T is merely the artifact coming from the loss of the magnetization due to the cancellation of the ferromagnetic signal of the film by the diamagnetism of the substrate. Reprinted from [34] by The Author(s) licensed under CC BY 4.0

magnetization with rectangular hysteresis loops in the thin films (Fig. 4.4e–g), in stark contrast to the magnetization curve of the CGT bulk crystal with no discernible hysteresis (Fig. 4.4d) as is the case for the previous studies even in flakes as thin as two monolayers [16, 18, 28, 32, 35]. The saturation magnetizations M_s of the thin film samples are about $2.4\mu_B$/Cr atom, almost irrespective of the thickness (see the inset of Fig. 4.4c). In contrast, the coercive fields increase with decreasing the thickness, which is contrary to the behavior for conventional ferromagnets with perpendicular magnetic anisotropy because the demagnetization field increases with decreasing thickness.

To reveal the peculiar behavior of the magnetic hysteresis loops, we measured the magnetization by applying a magnetic field along the in-plane direction ($B \perp c$) (black lines in Fig. 4.4d–g). The saturation magnetic field (indicated by a black arrow) increases with decreasing thickness, which is consistent with the enhancement of the

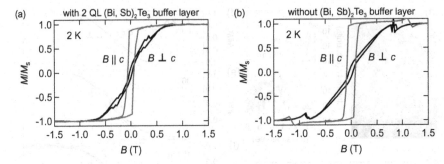

Fig. 4.5 Magnetization (M) curves normalized by the saturation magnetization (M_s) for 36-nm-thick $Cr_2Ge_2Te_6$ films with 2 QL $(Bi, Sb)_2Te_3$ buffer layer **a** and without $(Bi, Sb)_2Te_3$ buffer layer **b** grown under the identical condition of $P_{Ge}/P_{Cr} = 3.2$. Reprinted from [34] by The Author(s) licensed under CC BY 4.0

coercive fields if we consider a simple magnetization rotation reversal model. Upon the analogy of the perpendicular magnetic anisotropy as observed for Co/Pt films [37], the enhancement of the perpendicular magnetic anisotropy may be ascribed to the strong spin-orbit coupling in the adjacent BST layer. However, this scenario cannot account for the present case because another $Cr_2Ge_2Te_6$ film grown directly on an InP substrate also shows a similar rectangular hysteresis loop (Fig. 4.5). We speculate that the tendency of Ge deficiency is a major difference in MBE-grown films and the bulk single crystals possibly due to their different growth processes: non-equilibrium and equilibrium growth processes. Incidentally, similar thickness dependent enlargement of hysteresis loops is also observed in the exfoliation samples of ferromagnetic vdW $Fe_3Ge_2Te_2$ [38, 39], which may indicate a unique feature in two-dimensional ferromagnetism.

4.3 Large Anomalous Hall Effect in $Cr_2Ge_2Te_6/(Bi, Sb)_2Te_3$ Heterostructures

Now that the vdW CGT thin films are prepared by MBE, we fabricated a heterostructure consisting of CGT and BST (Fig. 4.6a). The structural characterization of the interface was carried out by STEM. Figure 4.6b displays the STEM image of a MBE-grown CGT(12 nm)/BST(9 nm)/CGT(12 nm) heterostructure, which exhibits abrupt interfaces with the ordered stacking orientation in favor of the hexagonal Te arrangements of CGT and BST. By performing Fourier transformation in the lateral direction of the image, the atomic distance of each layer is achieved as depicted in the right panel of Fig. 4.6b. Sharp changes of the lateral atomic distance at the interfaces reflect the weak epitaxial strain at the interfaces owing to vdW bonding. Furthermore, the XRD and magnetization also confirm the successful formation of the heterostructure (Fig. 4.7).

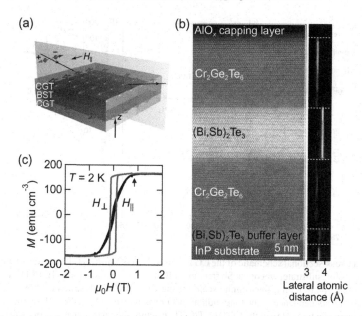

Fig. 4.6 **a** Schematic of the polarized neutron reflectometry (PNR) setup for the CGT/BST/CGT heterostructure. **b** Cross-sectional HAADF-STEM image of the CGT (12 nm)/BST (9 nm)/CGT (12 nm) heterostructure on a BST/InP substrate (left). The lateral atomic distance of each layer obtained by the Fourier transformation of the left image plotted along the growth direction (right). **c** Magnetization hysteresis loops under out-of-plane ($H_\perp$) and in-plane ($H_\parallel$) magnetic fields for the identical CGT(12 nm)/BST(9 nm)/CGT (12 nm) sample used in the STEM and PNR measurements. The black arrow represents the saturation field for the in-plane direction. Reprinted figure with permission from [40] Copyright 2019 by the American Physical Society

On the basis of the clean heterostructures, we have examined the interfacial magnetism of the CGT/BST/CGT sandwich heterostructure by depth-sensitive polarized neutron reflectometry (PNR). The PNR measurements, being qualitatively responsive to the in-plane magnetization, were conducted at 3 K under an in-plane magnetic field $\mu_0 H_\parallel = 1$ T (Fig. 4.8a), which is strong enough to fully align the magnetic moments to the field direction as confirmed by the magnetization hysteresis loops measured at 2 K (Fig. 4.6c). Figure 4.8c shows the x-ray and nonspin-flip specular PNR reflectivity curves R^+ and R^-, where $+(-)$ denotes the incident neutron spins parallel (antiparallel) to the direction of $H_\parallel$ as a function of the momentum transfer vector Q_z. The depth information of the in-plane saturated magnetization in the heterostructure is directly reflected in the spin-asymmetry ratio defined as $(R^+ - R^-)/(R^+ + R^-)$ (Fig. 4.8d). In addition, we combine an x-ray reflectivity (XRR) measurement at room temperature (Fig. 4.8c) to perform a model analysis for the structural parameters, including the thickness and roughness of each layer. The depth profile of the x-ray scattering length density (SLD) shown in (Fig. 4.8e), corresponding to the electron density distribution in the heterostructure, reflects the structural interface roughness. Notably, the root-mean-square roughness of all inter-

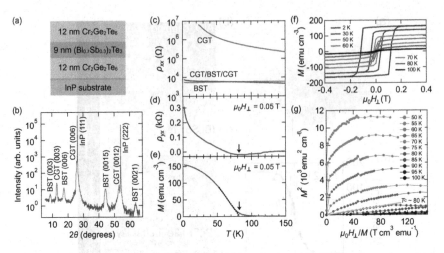

Fig. 4.7 **a** Cross-sectional schematic of the CGT(12 nm)/BST(9 nm)/CGT(12 nm) on a 1-nm-thick BST buffered InP substrate, where the Sb fraction of BST is $x = 0.3$, used for STEM/EDX and PNR studies. **b** XRD pattern on a logarithmic scale for the CGT/BST/CGT structure shown in **a**. **c–e** Temperature (T) dependence of the longitudinal sheet resistivity (ρ_{xx}) (**c**), the Hall resistivity (ρ_{yx}) **d** and the magnetization (M) **e** of the CGT/BST/CGT structure shown in **a**. In **c**, ρ_{xx} of a CGT single-layer ($t = 12$ nm) grown on a 1-nm-thick BST buffered InP substrate and a BST single-layer ($t = 9$ nm) directly grown on an InP substrate are also shown. Black arrows indicate the rising temperature of ρ_{yx} and M. **f** Perpendicular magnetic field ($\mu_0 H_\perp$) dependence of the magnetization (M) at various temperatures. **g** Arrott plot to determine the Curie temperature ($T_C \sim 80$ K). Reprinted figure with permission from [40] Copyright 2019 by the American Physical Society

faces in the SLD profiles is less than 1 nm, which is consistent with the STEM image shown in Fig. 4.6b. The structural parameters derived from the XRR fitted model were used to refine the PNR fitting analysis. Figure 4.8a and b display the magnetic SLD depth profiles based on the fitting results of R^+, R^-, and the spin-asymmetry ratio. The fitting analysis resolved the magnetizations of 152 ± 8 and 0 ± 20 emu/cm^3 for the CGT and BST layers, respectively. Although it is hard to precisely determine the induced magnetization within the BST layer due to the spatial broadening, it is reasonable to conclude that the induced magnetization in the BST layer is far smaller than the magnetization of the CGT layer.

We also perform the EDX measurements. As shown in Fig. 4.9, we confirm that almost no interdiffusion of atoms including Cr atoms, being consistent with the PNR result.

The magnetic proximity effect on the TI surface states can be inspected by magneto-transport measurements. The measurements were conducted with the sandwiched CGT/BST/CGT trilayers and the CGT/BST bilayers. Because of the high electric resistance of the CGT layer, its contribution to electrical transport is negligibly small (Fig. 4.7c). For the TI layer, instead of simple single-layered $(Bi_{1-x}Sb_x)_2Te_3$, we engineered a multilayer structure of $(Bi_{1-x}Sb_x)_2Te_3$ (2nm)/

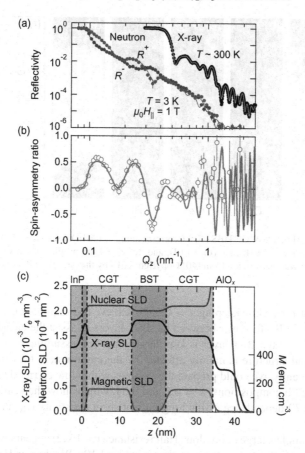

Fig. 4.8 **a** Measured (dots) and fitted (solid lines) reflectivity curves for the x-ray (black) and neutron of spin-up (R^+) (red) and spin-down (R^- (nlue) as a function of momentum transfer (Q_z) on a logarithmic scale. The error bars represent one standard deviation. **b** PNR spin-asymmetry ratio $(R^+ - R^-)/(R^+ + R^-)$ obtained from the experimental and fitted reflectivity curves in **a**. The error bars represent one standard deviation. **c** X-ray SLD (black) and neutron SLD divided into the nuclear (blue) and the magnetic (red) SLDs as a function of the distance from the InP substrate surface (z). The r_e in the unit of the x-ray SLD denots the classical electron radius of 2.82×10^{-15} m. For the magnetic SLD, the value of M corresponding to the neutron SLD is shown in the right ordinate. Reprinted figure with permission from [40] Copyright 2019 by the American Physical Society

Bi_2Te_3 (2 nm)/$(Bi_{1-x}Sb_x)_2Te_3$ (2 nm) (Fig. 4.10a) to suppress the effect of a charge transfer as discussed below.

In the CGT/BST/CGT heterostructures, the charge neutrality point is at a relatively small value of x ($0.3 < x < 0.4$) due to a possible hole transfer from CGT to BST (Fig. 4.11). According to an angle-resolved photoemission spectroscopy study on BST [41], small x causes the surface Dirac point to submerge below the bulk valence band. To approach the Dirac point with the charge neutrality condition, we need to

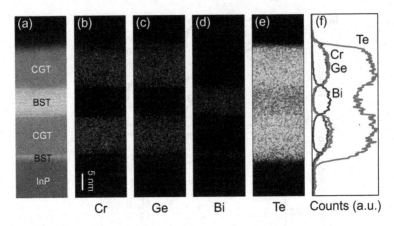

Fig. 4.9 **a** STEM image of CGT/BST/BST corresponding to the EDX scan area. **b–e** Distribution maps of each element, Cr (**b**), Ge (**c**), Bi (**d**), and Te (**e**). **f** Line profiles of Cr, Ge, Bi, and Te. Reprinted figure with permission from [40] Copyright 2019 by the American Physical Society

introduce electrons while keeping $x > 0.5$. To achieve this requirement, we inserted the electron-rich Bi_2Te_3 layer between the BST layers to assist the electron doping. The value of x in the BST layer is kept larger than 0.5, assuming that the surface electronic band structure is mainly affected by the environment near the interface [42]. Consequently, we could prepare the samples with a low carrier density at reasonably large Sb compositions, $x = 0.5$ and 0.64, which show the semiconducting temperature (T) dependence of the longitudinal sheet resistivity (ρ_{xx}) as shown in Fig. 4.10b.

In these samples, large anomalous Hall resistance ($> 1\,k\Omega$) appears with perpendicular anisotropic hysteresis loops as shown in Fig. 4.10c. We show in Fig. 4.10d the x dependence of the sheet carrier density (n_{2D}), the longitudinal sheet conductivity (σ_{xx}), and the Hall conductivity (σ_{xy}) at zero magnetic field as converted from ρ_{xx} and ρ_{yx}. The notable feature is that the σ_{xy} exceeds $0.2e^2/h$ in the most insulating sample ($x = 0.6$), where $\sigma_{xx} \sim 2e^2/h$ and $n_{2D} \sim 10^{12}$ cm^{-2}. The sheet carrier densities are estimated from the slope of Hall resistance above the saturation field. The carrier types are electrons for $x = 0.3$ and holes for $x = 0.64$, demonstrating that the Fermi level is systematically shifted from n to p type with increasing x (Fig. 4.10d).

These observations in Fig. 4.10 can be understood by an exchange gap opening in the dispersion relation of the TI surface state. When an exchange gap opens the surface of the TI, the Berry curvature is strongly enhanced near the band edge, resulting in the large σ_{xy}. Simultaneously, when the Fermi energy is tuned within or close to the exchange gap, σ_{xy} takes a maximum, while σ_{xx} takes a minimum. In the present study, we observe that σ_{xy} takes a maximum accompanied by a nearly minimum value of σ_{xx} in the sample with the low carrier density ($x = 0.6$) as expected. Also, the decrease in σ_{xy} and increase in σ_{xx} are observed as the carrier density is detuned from the optimum value. These carrier density dependences are consistent with the scenario

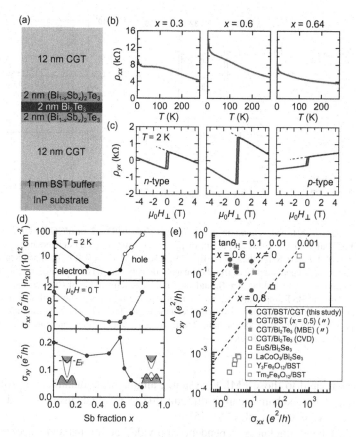

Fig. 4.10 a Schematic layout of CGT (12 nm)/(Bi$_{1-x}$Sb$_x$)$_2$Te$_3$ (2 nm)/Bi$_2$Te$_3$ (2 nm)/(Bi$_{1-x}$Sb$_x$)$_2$Te$_3$ (2 nm)/CGT (12 nm) heterostructure. **b, c,** Temperature (T) (out-of-plane magnetic field ($\mu_0H_\perp$)) dependence of the sheet resistivity (ρ_{xx}) in zero magnetic field **b** [the Hall resistivity (ρ_{yx}) at 2 K **c**] of CGT/(Bi$_{1-x}$Sb$_x$)$_2$Te$_3$/Bi$_2$Te$_3$/(Bi$_{1-x}$Sb$_x$)$_2$Te$_3$/CGT ($x = 0.3, 0.6, 0.64$) heterostructures. **d** Sb fraction (x) dependence of the sheet carrier density ($|n_{2D}|$) (top), the longitudinal sheet conductivity (σ_{xx}) (middle), and the Hall conductivity (σ_{xy}) (bottom) at 2 K. (Insets) Simplified schematics of band structures representing the different Fermi energies; the blue lines represent the surface state dispersion. **e** The anomalous Hall conductivity σ_{xy}^A as a function of the σ_{xx} with the use of the data shown **d**, comparing with other various FMI-TI heterostructures. The values of σ_{xy}^A and σ_{xx} are taken from the data obtained at the lowest temperature in the measurements (2–5 K) for the fully out-of-plane magnetized state. Reprinted figure with permission from [40] Copyright 2019 by the American Physical Society

described by the magnetic proximity effect. The increase of σ_{xy} accompanied by the decrease of σ_{xx} leads to an enhancement of the Hall angle $\theta_H = \tan^{-1}(\sigma_{xy}/\sigma_{xx})$, discriminating the anomalous Hall effect by extrinsic origins [43]. The obtained values of the σ_{xy} and the θ_H are dramatically increased in our samples compared to those reported in other FMI-TI systems (Fig. 4.10e). This trend suggests that the Fermi level is close to the exchange gap and/or that the exchange gap is large in our

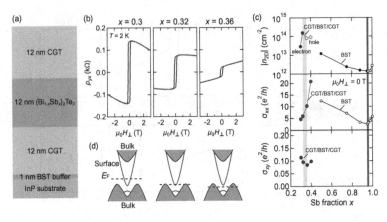

Fig. 4.11 **a** Cross-sectional schematic of the CGT(12 nm)/BST(12 nm)/CGT(12 nm) on a 1-nm-thick BST buffered InP substrate. **b** Out-of-plane magnetic field (B) dependence of the Hall resistivity (ρ_{yx}) at $T = 2$ K for the CGT/$(Bi_{1-x}Sb_x)_2Te_3$/CGT ($x = 0.3, 0.32, 0.36$) heterostructures. **c** Sb fraction (x) dependence of the sheet carrier density ($|n_{2D}|$) (top panel), the longitudinal sheet conductivity (σ_{xx}) (middle panel), and the Hall conductivity (sxy) (bottom panel) in the CGT/BST/CGT and BST single-layer films. The data for the BST single-layer films is an excerpt from [41]. **d** Schematics of the relationship between the Fermi-level (E_F) and the TI band structure for the respective films which are shown in **b**. Reprinted figure with permission from [40] Copyright 2019 by the American Physical Society

samples compared to those FMI-TI systems, although the quantitative estimation of the size of the exchange gap is difficult due to residual disorders and spatial inhomogeneities in the samples [44, 45].

The CGT-layer thickness dependence provides additional evidence that the observed anomalous Hall effect is induced by the magnetic proximity effect, excluding other possible origins arising from Cr diffusion into the BST layer. Figure 4.12a shows the T dependent magnetization ($M - T$) curves of four CGT/BST bilayers (inset of Fig. 4.12c) films with representative CGT-layer thicknesses which are evaluated by XRR (Fig. 4.13) under $\mu_0 H_\perp = 0.05$ T. Both the magnetization M and Curie temperature T_C decrease systematically with decreasing the CGT layer thickness t. As shown in Fig. 4.12b, the low-temperature values of σ_{xy} also decrease with decreasing t. In Fig. 4.12c, the t dependences of σ_{xy} at $T = 2$ K and the saturated magnetization of CGT M_s are plotted together. The agreement in t dependences of σ_{xy} and M_s indicate that σ_{xy} is almost proportional to M_s. This observation indicates that the exchange gap on the TI surface state can be tuned by the magnetization of the CGT layer, directly pointing to the proximity-coupling origin of the anomalous Hall effect. The decrease in M_s in the range of $t < 2$ nm is attributed perhaps to the dimensionality effect of the 2D ferromagnetic CGT layer [16]. In contrast to σ_{xy}, σ_{xx} is almost constant for the variation of t across $t \sim 2$ nm (Fig. 4.12c). This constancy of σ_{xx} suggests that the Fermi energy and the scattering time are not largely affected by the thickness of the CGT layer.

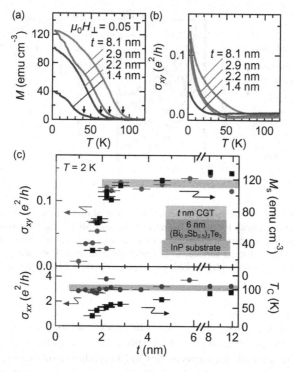

Fig. 4.12 a $M - T$ and **b** $\sigma_{xy} - T$ curves measured under a field cooling under $\mu_0 H_\perp = 0.05$ T in BST/CGT ($t = 1.4, 2.2, 2.9$, and 8.1 nm) bilayer structures. The black arrows indicate the Curie temperature T_C to highlight the changes against t. **c** CGT thickness t dependence of the σ_{xy} (red closed circles, left red axis) and the spontaneous magnetization M_s (black closed squares, right black axis) at 2 K under zero magnetic field (top) and σ_{xx} at 2 K (blued closed circles, left blue axis) and T_C (black closed squares, right black axis) (bottom). The inset shows a schematic layout of $(Bi_{0.5}Sb_{0.5})_2Te_3$(6 nm)/CGT(t nm) bilayer structure. Reprinted figure with permission from [40] Copyright 2019 by the American Physical Society

On the basis of the above our results, we discuss the possible mechanism of the formation of the exchange gap the CGT/BST interface. One conceivable scenario would be the induction of magnetization in the TI layer by the adjacent FMI layer as discussed in earlier works [3–10]. However, this scenario is unlikely applicable to the present case. At the interfaces of EuS/Bi_2Se_3 and EuS/BST, large magnetizations of about 270 and 160 emu/cm³, respectively, have been reported to appear [11, 12]. By contrast, in the present study, the magnetization of the CGT layer is already smaller than these values. Therefore, the induced magnetization, if any, in the BST layer of the present CGT/BST heterostructure would be much smaller than that reported for the EuS-based heterostructures. Despite the small induced magnetization, the present transport measurements have revealed that the σ_{xy} and θ_H are much enhanced in the CGT/BST system. One other possible scenario to understand these observations is the formation of the exchange gap by penetration of the TI surface state wave function

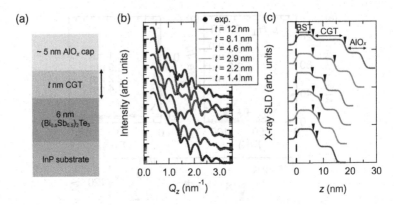

Fig. 4.13 **a** Schematic of the sample structure. **b** Measured (dots) and fitted (solid lines) XRR curves on a logarithmic scale for the BST(6 nm)/CGT(t nm)/AlO$_x$(5 nm) structures with various CGT thicknesses as a function of momentum transfer (Q_z). **c** X-ray SLDs as a function of the distance from the InP substrate surface, z. Triangles indicate the bottom and the top surface of CGT layers. Reprinted figure with permission from [40] Copyright 2019 by the American Physical Society

into the FMI layer. In this picture, even if the interfacial magnetization in the BST layer is small, the penetration part of the surface state wavefucntions can strongly interact with the magnetic moments in the CGT to produce a sizable exchange gap. A first-principles calculation work indicates the formation of a large exchange gap in the Te-based heterostructure MnBi$_2$Te$_4$/Bi$_2$Te$_3$ based on the wavefunction penetration mechanism [15].

4.4 Signature of Quantum Anomalous Hall Effect in Cr$_2$Si$_2$Te$_6$/(Bi, Sb)$_2$Te$_3$ Heterostructures

Having established the proximity-driven anomalous Hall effect in the combination between Te-based BST and CGT, we search an alternative FMI to enhance the anomalous Hall effect and to realize the QAH effect. As discussed in the previous section, the possible hole transfer from the Cr$_2$Ge$_2$Te$_6$ layer to (Bi, Sb)$_2$Te$_3$ layer makes it difficult to tune E_F to the Dirac point because the Dirac point for Bi-rich (Bi, Sb)$_2$Te$_3$ can be buried in the bulk valence bands [41]. In this section, we try to reduce the charge transfer in heterostructures based on (Bi, Sb)$_2$Te$_3$/vdW ferromagnetic insulator structure. We chose Cr$_2$Si$_2$Te$_6$ (CST) [46, 47] as an alternative FMI layer; the reasons are as follows: (1) The crystal structure of CST (Fig. 4.14b) is the same as that of CGT, which possibly enables us to grow clean heterostructures with BST as well as the BST/CGT heterostructures on the basis of vdW epitaxy. (2) CST has more insulating properties than CGT because of the larger band gap ~ 0.6 eV than CGT

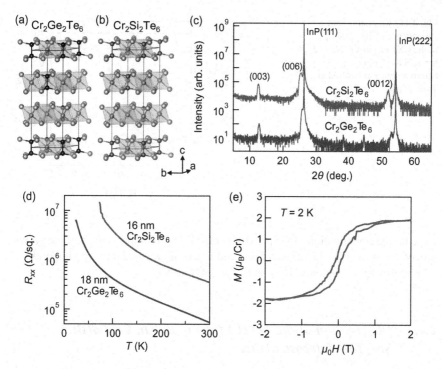

Fig. 4.14 **a**, **b** Schematics of crystal structures for $Cr_2Ge_2Te_6$ (CGT) (**a**) and $Cr_2Si_2Te_6$ (CST) (**b**). **c** XRD patterns for the MBE-grown CGT and CST thin films. **d** Temperature dependence of longitudinal resistance R_{xx} for the CST and CGT thin films. **e** M-H curve for the CST thin film measured at $T = 2$ K

(~ 0.4 eV) [28, 47, 48]. Thus, we expect a reduced charge transfer while keeping the efficient proximity coupling.

MBE-growth of the CST thin films were performed in a similar manner as CGT. As shown in Fig. 4.14c, the XRD pattern of the CST film confirms the same crystal structure as CGT. The slightly elongated c-axis length of the CST (2.11 nm) film compared to the CGT films (2.09 nm) are also consistent with the bulk values [46]. The improved insulating property in CST is affirmed by the transport measurement (Fig. 4.14d), where the sheet resistivity of the CST film ($\sim 0.6\Omega cm$ at room temperature) is roughly one order of magnitude higher than that of the CGT film (~ 0.09 Ωcm at room temperature) while the ferromagnetic property is also kept as seen in Fig. 4.14e.

Now that the CST FMI layer is successfully grown, we create a heterostructure of CST (15 nm)/ BST (12 nm)/CST (15 nm). We show the Hall measurement at the lowest temperature (60 mK) (Fig. 4.15). The anomalous Hall effect is significantly enhanced compared to CGT-based heterostructure, where the spontaneous anomalous Hall resistance is near the quantized value of $\sim 0.8h/e^2$. Thus, we have confirmed the consideration of charge transfers between TI and FMI layers is a crucial

Fig. 4.15 Magnetic field
dependence of ρ_{xx} and ρ_{yx}
in the CST (15 nm)/BST (12
nm)/CST (15 nm)
heterostructure measured at
$T = 60$ mK

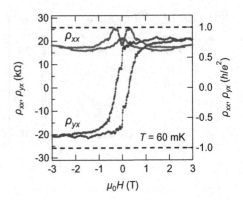

issue for the investigation of the proximity effect. Future improvement of the growth procedure of the CST layer would proceed to the full quantization and hopefully to even higher-temperature realization of the QAH effect.

4.4.1 Observation of the QAH Effect in a (Zn, Cr)Te/(Bi, Sb)$_2$Te$_3$ Heterostructure

We further applied the strategy for realizing the efficient magnetic proximity coupling based on Te-based heterostructures to a diluted magnetic semiconductor (Zn, Cr)Te (ZCT), which surprisingly has both ferromagnetic and insulating properties [49]. With increasing the density of Cr atoms in ZnTe, the proximity-induced anomalous Hall angle is dramatically enhanced. Finally, at the highest concentration of Cr (17%) in the ZCT layer without segregation of Cr atoms, the QAH effect is successfully observed up to 100 mK. This demonstration paves a route to design the functionalities with the use of the magnetic layers degrees of freedom.

4.5 Current-Induced Magnetization Switching of Ferromagnetic Insulator Through Topological Surface State

The electrical magnetization switching of magnetic materials is one of the most important spintronic functionalities for low-energy consumption logic and memory device applications [51]. Current-induced spin-orbit torque (SOT) generation in large spin-orbit coupling materials/magnetic materials junctions has provided an efficient way to control the magnetization as compared to the conventional spin-transfer torque generation in spin-valves and magnetic tunnel junctions [52, 53].

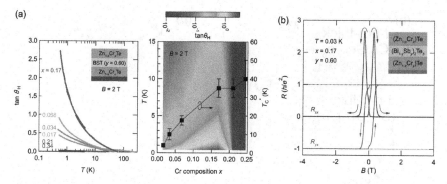

Fig. 4.16 **a** Temperature dependence (left) and the color map (right) of the tangent Hall angle $\tan \theta_H$ for the ZCT/BST/ZCT heterostructure with various Cr composition x under a magnetic field of $B = 2$ T. Black closed squares represent T_C. **b** Magnetic field dependence of ρ_{yx} and ρ_{xx} at $T = 30$ mK. Reproduced from Ref. [50]

For example, in non-magnetic heavy metals, spin currents converted from charge currents exert SOT on the proximate magnetic layers through the spin-Hall effect [54–56]. Moreover, TIs are a class of materials possessing spin-polarized metallic surface states with insulting bulk states, which is expected as an ideal SOT generator because the spin-momentum locked surface states accumulate the spin at the surface via the Rashba-Edelstein effect. The charge-spin conversion efficiency is one or two orders of magnitude larger than those in heavy metals as exemplified by spin-torque ferromagnetic resonance [57–59] or spin pumping measurements [60]. Furthermore, by using damping-like SOT, deterministic perpendicular magnetization switching has been reported in magnetically-doped TI heterostructures [61, 62] and TI/magnet heterostructures even above room temperature [63–66]. However, in the greater part of experiments on TI/magnet heterostructures, magnets possess far lower resistivity than TI, then most of the electrical current is shunted through the metallic magnet layers, leading to the loss of the spin torque generation.

Using insulators for the adjacent magnetic layers, since charge currents only flow in the TI surface state, the magnetization switching can be more energy efficient [67, 68]. Such attempts have been demonstrated in heavy metal systems, such as Pt/Tm$_3$Fe$_5$O$_{12}$ (TIG) [69] and W/TIG [70]. However, in magnetic insulator/TI junctions, reports on current-induced magnetization switching have been limited to a few examples [71], whereas various spintronic properties including microwave/current-driven spin dynamics have been studied in TIG/TI and YIG/TI bilayers [67, 68, 72]. Currently, the contribution of the surface state against the bulks states to the magnetization switching remains elusive.

Here, we realize the current-induced SOT switching in the BST/CGT bilayer structures (Fig. 4.17a), where the perpendicular magnetization opens an exchange gap in the surface state through magnetic proximity effect (Fig. 4.17b). Owing to the large anomalous Hall effect in the BST/CGT heterostructures, which is much larger than the previously reported oxide-based FMI/TI heterostructures, the itinerant spins

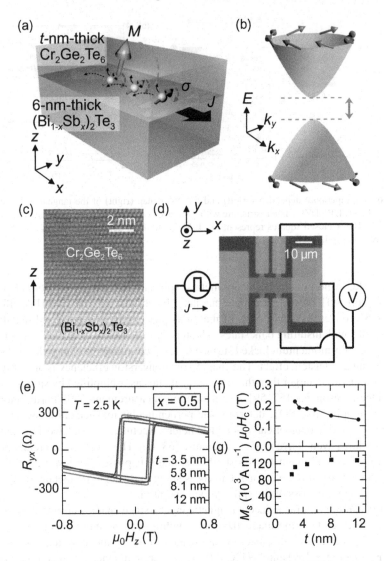

Fig. 4.17 **a** Schematic illustration of magnetization switching in a BST/CGT bilayer structure. Out-of-plane and in-plane components of the spin (σ) direction in TI surface states generate anomalous Hall effect and spin-orbit torque (SOT) on CGT, respectively. **b** Schematic of TI surface band dispersion with an exchange gap induced by proximity coupling. **c** Cross-sectional STEM image of a BST/CGT heterostructure. **d** Optical microscope image of a Hall bar device with an illustration of the measurement setup. **e** Out-of-plane magnetic field ($\mu_0 H_z$) dependence of the Hall resistance (R_{yx}) for $(Bi_{0.5}Sb_{0.5})_2Te_3$ (6 nm)/CGT bilayers with various CGT thickness ($t = 3.5, 5.8, 8.1$, and 12 nm) at 2.5 K. **f, g** t dependence of the coercive field ($\mu_0 H_c$) and the spontaneous magnetization (M_s)

of the TI surface state is expected to be efficiently transferred to the localized spins in the CGT layer although the microscopic origin of the strong damping-like SOT due to the Rashba-Edelstein effect is still under debate [53]. In the following, we show the CGT thickness dependence of transport properties and current-induced magnetization reversal, consistent with SOT generation at the interface of the BST/CGT. We also examine the E_F dependence by tuning the Bi/Sb composition (x), identifying that the SOT generation is dominated by the TI surface states compared to the bulk states.

Figure 4.17c shows a cross-sectional high-angle annular dark field scanning transmission image for a BST/CGT heterostructure, representing a highly sharp interface. For the sake of magnetization switching experiments, we fabricated Hall bar devices with a width of 10 μm and a length of 30 μm as shown in Fig. 4.17d. There, to minimize the heating the samples, we injected current pulses up to about $J = 6$ mA which corresponds to the current density of $j = 1 \times 10^7$ A cm^{-2}, considering that the resistivity of CGT films ($> 1\ \Omega$ cm at low temperatures) is much higher than that of BST films (< 10 mΩ cm). To determine the final state of the magnetization in the perpendicular SOT switching, we applied in-plane magnetic field ($H_x = 0.1$ T). After each pulse injection, we read Hall resistance (R_{yx}) originating from the anomalous Hall effect by small excitation currents of 10 μA ($= 1.67 \times 10^4$ A cm^{-2}) to ignore the sample heating. All the measurements were performed at 2.5 K which is fairly lower than the Curie temperature of CGT ($T_C \sim 80$ K). In Fig. 4.17e, we show the anomalous Hall effect of several 6-nm-thick BST ($x = 0.5$) / CGT bilayers with various t under out-of-plane magnetic fields (H_z). The perpendicular anisotropic hysteresis loops with perpendicular remanence are consistent with the ferromagnetic property of MBE-grown CGT films. Owing to the magnetic proximity effect that the TI surface state interacts with the Cr ions in the CGT layers, the anomalous Hall resistivity (R_{yx}^{AH}) (Fig. 4.17f) and the carrier density as judged from the normal Hall coefficient are highly constant against the variation of t. For magnetic properties of the CGT layers, the coercive field (H_c) slightly decreases with increasing t as summarized in Fig. 4.17e. On the other hand, as shown in Fig. 4.17g, the spontaneous magnetization per volume (M_s) is almost constant to about 120 kA/m when $t > 2.9$ nm whereas a dimensional effect of the thinned CGT layer appears when $t < 2.9$ nm (Fig. 4.17g) [16]. Hence, we focus on $t > 2.9$ nm in the following experiments.

Figure 4.18a shows the observation of current-induced magnetization switching under in the identical samples shown in Fig. 4.17e. Firstly, we focus on the result for the $t = 3.5$ nm sample. The current pulse changes the Hall resistance ΔR_{yx}. The ΔR_{yx} value approximately corresponds to the value of R_{yx}^{AH} ($\Delta R_{yx}/R_{yx}^{AH} \sim 0.8$), indicating almost all the magnetization of the CGT layers is switched between up and down directions. In addition, when the direction of H_x is reversed, the switching direction is also reversed, which is typical behavior of the damping-like SOT perpendicular magnetization switching [56]. Moreover, this switching direction is consistent with the spin accumulation from the spin-momentum locked BST surface states [73]. In this case, the accumulated spins as described by $\sigma = +\sigma_y \hat{y}$ can generate the anti-damping SOT $\tau_{SO} = -M \times H_{SO} = M \times (M \times \sigma)$ or the effective field

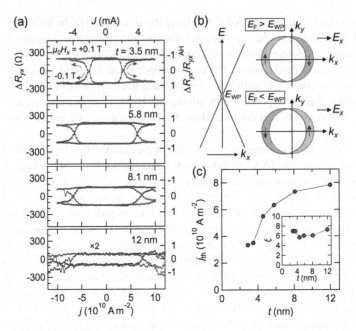

Fig. 4.18 a Variation of Hall resistance (ΔR_{yx}, left ordinate) by current (J, upper abscissa) and the corresponding current density (j, lower abscissa) pulse injection under in-plane magnetic fields $\mu_0 H_x = +0.1$ T (red) and -0.1 T (blue) at 2.5 K in the $(Bi_{0.5}Sb_{0.5})_2Te_3$ (6 nm)/$Cr_2Ge_2Te_6$ (t nm) samples. Right ordinate indicates the switching ratio defined as $\Delta R_{yx}/R_{yx}^{AH}$. **b** Illustration of the spin-accumulation due to the Rashba-Edelstein effect from the spin-momentum-locked TI Fermi surface. **c** t dependence of the threshold current density (jth). The inset of **c** represents t dependence of the switching ratio defined as $\Delta R_{yx}/R_{yx}^{AH}$ and a dimensionless quantity defined as $\xi = (2e/\hbar)(M_s t \mu_0 H_c / jth)$, respectively

$H_{SO} = -M \times \sigma = x$ direction, which is maintained regardless of the Fermi energy against the Dirac point as shown in Fig. 4.18b.

To evaluate the efficiency of the magnetization switching, we define the threshold current density (jth) where ΔR_{yx} takes zero. For the $t = 3.5$ nm sample, jth is as small as 3.5×10^{10} A m^{-2}. This is typically one order of magnitude smaller value than those of heavy metal systems [54, 56] and is comparable to those of magnet/TI and magnetically doped TI systems [61–66, 71]. We further investigate thicker-CGT heterostructures. As shown in Fig. 4.18b, jth becomes large with increasing thickness t. This is mainly because the magnetization per sample area ($M_s t$) linearly increases. To look into the t dependence, we plot jth as a function of t in Fig. 4.18c. With increasing t, jth increases but does not follow the t-linear relation. We find that the collapse of t-linear relation can be explained by the slight change of H_c with the variation of t as shown in Fig. 4.18f. To include the effect of magnetic properties, such as H_c and M_s, we define a dimensionless quantity defined as $\xi = (2e/\hbar)(M_s t \mu_0 H_c / jth)$ (Ref. [53]), where e is the elementary charge and μ_0 is the vacuum permeability. In the inset of Fig. 4.18c, we plot ξ as a function of t, which

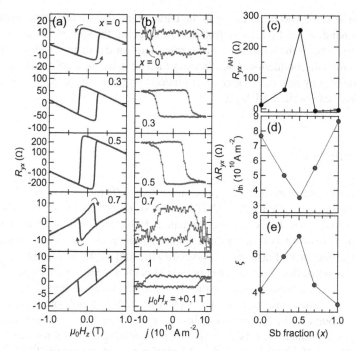

Fig. 4.19 a Out-of-plane magnetic field ($\mu_0 H_z$) dependence of the Hall resistance (R_{yx}) for $(Bi_{1-x}Sb_x)_2Te_3$(6 nm)/CGT(3.5 nm) heterostructures with various x (= 0, 0.3, 0.5, 0.7, 1) at 2.5 K. **b** Variation of Hall resistance (ΔR_{yx}) as a function of the injected pulse current density (j) injection under in-plane magnetic fields $\mu_0 H_x = +0.1$ T at 2.5 K. **c–e** Sb fraction (x) dependence of the switching ratio $\Delta R_{yx}/R_{yx}^{AH}$ (**c**), the threshold current density jth (**d**), and the dimensionless quantity $\xi = (2e/\hbar)(M_s t \mu_0 H_c/j\text{th})$ (**e**)

roughly becomes constant against t. Eventually, we can use ξ as an efficiency index for the interpretation of E_F dependence as below.

Figure 4.19a and c shows the variation of the Hall resistance traces and R_{yx}^{AH}, respectively, in $(Bi_{1-x}Sb_x)_2Te_3$(6 nm)/CGT(3.5 nm) heterostructures by varying x. With increasing x up to 0.5, R_{yx}^{AH} increases because E_F is approaching the gapped surface Dirac point. For $x = 0.7$ and 1, the sign of R_{yx}^{AH} becomes negative. The similar sign change has been reported in BST/Cr-doped BST heterostructures [74]; it has been pointed out that the negative R_{yx}^{AH} can occur by considering opposite sign of orbital dependent exchange coupling [75], where a magnetic gap opening at the Rashba-splitting bulk valence bands (Te $6p$ orbitals) originating from broken inversion symmetry [74]. We compare the traces with current-induced magnetization switching (Fig. 4.19b). The switching direction and the sign of anomalous Hall effect coincide with each other, indicating the switching direction is static against changing x, being consistent with the surface state dominated magnetization switching (Fig. 4.18b). Importantly, jth varies dramatically (Fig. 4.19d). Especially, for the $x = 0.5$ sample showing the largest R_{yx}^{AH}, jth becomes the minimum. This indicates that when E_F becomes closer to the Dirac point, the jth becomes smaller. On the other hand, when E_F is away from the Dirac point, the spin-polarization can be smaller because of the existence of a warping term in the surface states and/or the concomitant bulk conduction. The efficiency factor ξ is also in accord with the change of jth, suggesting that the change of jth indeed originates from the E_F change for the surface state of the BST layer.

4.6 Conclusion

We have investigated the magnetic proximity effect on TI by using van der Waals ferromagnetic insulators $Cr_2Ge_2Te_6$ (CGT) and $Cr_2Si_2Te_6$ (CST). In the CGT/BST/CGT heterostructures, we observe a large anomalous Hall effect, in which the anomalous Hall angle is two-orders of magnitude larger than the previous reports on magnetic proximity effect in TI. Furthermore, we clarify the film-depth profile of the structure and the magnetization via x-ray and polarized neutron reflectometry, identifying a large scale ($\sim$ mm) homogeneity of the chemical structure and the magnetization at the interfaces. These results suggest a mechanism that the topological surface states efficiently couple to the ferromagnetic insulator via the surface states penetrating the ferromagnetic insulator layers. By using more insulating CST compared with CGT we are able to effectively control the chemical potential toward the Dirac point of the surface state owing to reduced charge transfer compared to the BST/CGT heterostructures. Thus, the CST/BST/CST heterostructures has shown the signature of the QAH effect, where the Hall conductivity tentatively enhances up to $\sim 0.8e^2/h$. Future improvement of the growth condition would lead to the full quantization and hopefully to even higher-temperature realization of the QAH effect. Furthermore, the current-induced spin-orbit torque by the spin-momentum locked topological surface state can almost completely switch the magnetization of CGT. This would open a way for electrically switching the chiral edge states of the QAH effect.

References

1. W. Luo, X.L. Qi, Phys. Rev. B **87**(8), 085431 (2013)
2. S. Eremeev, V. Men'Shov, V. Tugushev, P.M. Echenique, E.V. Chulkov, Phys. Rev. B **88**(14), 144430 (2013)
3. P. Wei, F. Katmis, B.A. Assaf, H. Steinberg, P. Jarillo-Herrero, D. Heiman, J.S. Moodera, Phys. Rev. Lett. **110**(18), 186807 (2013)
4. Q.I. Yang, M. Dolev, L. Zhang, J. Zhao, A.D. Fried, E. Schemm, M. Liu, A. Palevski, A.F. Marshall, S.H. Risbud et al., Phys. Rev. B **88**(8), 081407 (2013)
5. A. Kandala, A. Richardella, D. Rench, D. Zhang, T. Flanagan, N. Samarth, Appl. Phys. Lett. **103**(20), 202409 (2013)
6. M. Lang, M. Montazeri, M.C. Onbasli, X. Kou, Y. Fan, P. Upadhyaya, K. Yao, F. Liu, Y. Jiang, W. Jiang et al., Nano Lett. **14**(6), 3459 (2014)

7. L. Alegria, H. Ji, N. Yao, J. Clarke, R.J. Cava, J.R. Petta, Appl. Phys. Lett. **105**(5), 053512 (2014)
8. Z. Jiang, C.Z. Chang, C. Tang, P. Wei, J.S. Moodera, J. Shi, Nano Lett. **15**(9), 5835 (2015)
9. C. Tang, C.Z. Chang, G. Zhao, Y. Liu, Z. Jiang, C.X. Liu, M.R. McCartney, D.J. Smith, T. Chen, J.S. Moodera et al., Sci. Adv. **3**(6), e1700307 (2017)
10. S. Zhu, D. Meng, G. Liang, G. Shi, P. Zhao, P. Cheng, Y. Li, X. Zhai, Y. Lu, L. Chen et al., Nanoscale **10**(21), 10041 (2018)
11. F. Katmis, V. Lauter, F.S. Nogueira, B.A. Assaf, M.E. Jamer, P. Wei, B. Satpati, J.W. Freeland, I. Eremin, D. Heiman et al., Nature **533**(7604), 513 (2016)
12. M. Li, Q. Song, W. Zhao, J.A. Garlow, T.H. Liu, L. Wu, Y. Zhu, J.S. Moodera, M.H. Chan, G. Chen et al., Phys. Rev. B **96**(20), 201301 (2017)
13. T. Hirahara, S.V. Eremeev, T. Shirasawa, Y. Okuyama, T. Kubo, R. Nakanishi, R. Akiyama, A. Takayama, T. Hajiri, S.i. Ideta, et al., Nano Lett. **17**(6), 3493 (2017)
14. V. Men'shov, V. Tugushev, S. Eremeev, P.M. Echenique, E.V. Chulkov, Phys. Rev. B **88**(22), 224401 (2013)
15. M.M. Otrokov, T.V. Menshchikova, M.G. Vergniory, I.P. Rusinov, A.Y. Vyazovskaya, Y.M. Koroteev, G. Bihlmayer, A. Ernst, P.M. Echenique, A. Arnau, et al., 2D Mater. **4**(2), 025082 (2017)
16. C. Gong, L. Li, Z. Li, H. Ji, A. Stern, Y. Xia, T. Cao, W. Bao, C. Wang, Y. Wang et al., Nature **546**(7657), 265 (2017)
17. B. Huang, G. Clark, E. Navarro-Moratalla, D.R. Klein, R. Cheng, K.L. Seyler, D. Zhong, E. Schmidgall, M.A. McGuire, D.H. Cobden et al., Nature **546**(7657), 270 (2017)
18. W. Xing, Y. Chen, P.M. Odenthal, X. Zhang, W. Yuan, T. Su, Q. Song, T. Wang, J. Zhong, S. Jia, et al., 2D Mater. **4**(2), 024009 (2017)
19. S. Jiang, J. Shan, K.F. Mak, Nat. Mater. **17**(5), 406 (2018)
20. B. Huang, G. Clark, D.R. Klein, D. MacNeill, E. Navarro-Moratalla, K.L. Seyler, N. Wilson, M.A. McGuire, D.H. Cobden, D. Xiao et al., Nat. Nanotech. **13**(7), 544 (2018)
21. S. Jiang, L. Li, Z. Wang, K.F. Mak, J. Shan, Nat. Nanotech. **13**(7), 549 (2018)
22. T. Song, X. Cai, M.W.Y. Tu, X. Zhang, B. Huang, N.P. Wilson, K.L. Seyler, L. Zhu, T. Taniguchi, K. Watanabe et al., Science **360**(6394), 1214 (2018)
23. D.R. Klein, D. MacNeill, J.L. Lado, D. Soriano, E. Navarro-Moratalla, K. Watanabe, T. Taniguchi, S. Manni, P. Canfield, J. Fernández-Rossier et al., Science **360**(6394), 1218 (2018)
24. D. Zhong, K.L. Seyler, X. Linpeng, R. Cheng, N. Sivadas, B. Huang, E. Schmidgall, T. Taniguchi, K. Watanabe, M.A. McGuire et al., Sci. Adv. **3**(5), e1603113 (2017)
25. A.K. Geim, I.V. Grigorieva, Nature **499**(7459), 419 (2013)
26. K. Novoselov, A. Mishchenko, A. Carvalho, A.C. Neto, Science **353**(6298), aac9439 (2016)
27. D. Pesin, A.H. MacDonald, Nat. Mater. **11**(5), 409 (2012)
28. H. Ji, R. Stokes, L. Alegria, E. Blomberg, M. Tanatar, A. Reijnders, L.M. Schoop, T. Liang, R. Prozorov, K. Burch et al., J. Appl. Phys. **114**(11), 114907 (2013)
29. T. Hashimoto, K. Hoya, M. Yamaguchi, I. Ichitsubo, J. Phys. Soc. Jpn. **31**(3), 679 (1971)
30. T. Hamasaki, T. Hashimoto, Y. Yamaguchi, H. Watanabe, Solid State Commun. **16**(7), 895 (1975)
31. A. Roy, S. Guchhait, R. Dey, T. Pramanik, C.C. Hsieh, A. Rai, S.K. Banerjee, ACS Nano **9**(4), 3772 (2015)
32. V. Carteaux, D. Brunet, G. Ouvrard, G. Andre, J. Phys.: Condens. Matter **7**(1), 69 (1995)
33. Y. Zhang, K. He, C.Z. Chang, C.L. Song, L.L. Wang, X. Chen, J.F. Jia, Z. Fang, X. Dai, W.Y. Shan et al., Nat. Phys. **6**(8), 584 (2010)
34. M. Mogi, A. Tsukazaki, Y. Kaneko, R. Yoshimi, K. Takahashi, M. Kawasaki, Y. Tokura, APL Mater. **6**(9), 091104 (2018)
35. X. Zhang, Y. Zhao, Q. Song, S. Jia, J. Shi, W. Han, Jpn. J. Appl. Phys. **55**(3), 033001 (2016)
36. R.J. Cava, H. Ji, M.K. Fuccillo, Q.D. Gibson, Y.S. Hor, J. Mater. Chem. C **1**(19), 3176 (2013)
37. M. Johnson, P. Bloemen, F. Den Broeder, J. De Vries, Rep. Prog. Phys. **59**(11), 1409 (1996)
38. C. Tan, J. Lee, S.G. Jung, T. Park, S. Albarakati, J. Partridge, M.R. Field, D.G. McCulloch, L. Wang, C. Lee, Nat. Commun. **9**(1), 1554 (2018)

39. Z. Fei, B. Huang, P. Malinowski, W. Wang, T. Song, J. Sanchez, W. Yao, D. Xiao, X. Zhu, A.F. May et al., Nat. Mater. **17**(9), 778 (2018)
40. M. Mogi, T. Nakajima, V. Ukleev, A. Tsukazaki, R. Yoshimi, M. Kawamura, K.S. Takahashi, T. Hanashima, K. Kakurai, T.h. Arima, et al., Phys. Rev. Lett. **123**(1), 016804 (2019)
41. J. Zhang, C.Z. Chang, Z. Zhang, J. Wen, X. Feng, K. Li, M. Liu, K. He, L. Wang, X. Chen et al., Nat. Commun. **2**, 574 (2011)
42. C.Z. Chang, P. Tang, X. Feng, K. Li, X.C. Ma, W. Duan, K. He, Q.K. Xue, Phys. Rev. Lett. **115**, 136801 (2015) https://doi.org/10.1103/PhysRevLett.115.136801. https://link.aps.org/doi/10.1103/PhysRevLett.115.136801
43. N. Nagaosa, J. Sinova, S. Onoda, A.H. MacDonald, N.P. Ong, Rev. Mod. Phys. **82**(2), 1539 (2010)
44. B. Skinner, T. Chen, B.I. Shklovskii, Phys. Rev. Lett. **109**, 176801 (2012) https://doi.org/10.1103/PhysRevLett.109.176801. https://link.aps.org/doi/10.1103/PhysRevLett.109.176801
45. N.A. Sinitsyn, J.E. Hill, H. Min, J. Sinova, A.H. MacDonald, Phys. Rev. Lett. **97**, 106804 (2006). https://doi.org/10.1103/PhysRevLett.97.106804. https://link.aps.org/doi/10.1103/PhysRevLett.97.106804
46. G. Ouvrard, E. Sandre, R. Brec, J. Solid State Chem. **73**(1), 27 (1988)
47. L. Casto, A. Clune, M. Yokosuk, J. Musfeldt, T. Williams, H. Zhuang, M.W. Lin, K. Xiao, R. Hennig, B. Sales et al., APL Mater. **3**(4), 041515 (2015)
48. Y.F. Li, W. Wang, W. Guo, C.Y. Gu, H.Y. Sun, L. He, J. Zhou, Z.B. Gu, Y.F. Nie, X.Q. Pan, Phys. Rev. B **98**, 125127 (2018). https://doi.org/10.1103/PhysRevB.98.125127. https://link.aps.org/doi/10.1103/PhysRevB.98.125127
49. H. Saito, W. Zaets, S. Yamagata, Y. Suzuki, K. Ando, J. Appl. Phys. **91**(10), 8085 (2002)
50. R. Watanabe, R. Yoshimi, M. Kawamura, M. Mogi, A. Tsukazaki, X. Yu, K. Nakajima, K.S. Takahashi, M. Kawasaki, Y. Tokura, Appl. Phys. Lett. **115**(10), 102403 (2019)
51. I. Žutić , J. Fabian, S. Das Sarma, Rev. Mod. Phys. **76**, 323 (2004). https://doi.org/10.1103/RevModPhys.76.323. https://link.aps.org/doi/10.1103/RevModPhys.76.323
52. A. Soumyanarayanan, N. Reyren, A. Fert, C. Panagopoulos, Nature **539**(7630), 509 (2016)
53. A. Manchon, J. Železný, I.M. Miron, T. Jungwirth, J. Sinova, A. Thiaville, K. Garello, P. Gambardella, Rev. Mod. Phys. **91**, 035004 (2019). https://doi.org/10.1103/RevModPhys.91.035004. https://link.aps.org/doi/10.1103/RevModPhys.91.035004
54. I.M. Miron, K. Garello, G. Gaudin, P.J. Zermatten, M.V. Costache, S. Auffret, S. Bandiera, B. Rodmacq, A. Schuhl, P. Gambardella, Nature **476**(7359), 189 (2011)
55. L. Liu, C.F. Pai, Y. Li, H. Tseng, D. Ralph, R. Buhrman, Science **336**(6081), 555 (2012)
56. L. Liu, O. Lee, T. Gudmundsen, D. Ralph, R. Buhrman, Phys. Rev. Lett. **109**(9), 096602 (2012)
57. A. Mellnik, J. Lee, A. Richardella, J. Grab, P. Mintun, M.H. Fischer, A. Vaezi, A. Manchon, E.A. Kim, N. Samarth et al., Nature **511**(7510), 449 (2014)
58. Y. Wang, P. Deorani, K. Banerjee, N. Koirala, M. Brahlek, S. Oh, H. Yang, Phys. Rev. Lett. **114**, 257202 (2015). https://doi.org/10.1103/PhysRevLett.114.257202. https://link.aps.org/doi/10.1103/PhysRevLett.114.257202
59. K. Kondou, R. Yoshimi, A. Tsukazaki, Y. Fukuma, J. Matsuno, K. Takahashi, M. Kawasaki, Y. Tokura, Y. Otani, Nat. Phys. **12**(11), 1027 (2016)
60. Y. Shiomi, K. Nomura, Y. Kajiwara, K. Eto, M. Novak, K. Segawa, Y. Ando, E. Saitoh, Phys. Rev. Lett. **113**(19), 196601 (2014)
61. Y. Fan, P. Upadhyaya, X. Kou, M. Lang, S. Takei, Z. Wang, J. Tang, L. He, L.T. Chang, M. Montazeri et al., Nat. Mater. **13**(7), 699 (2014)
62. K. Yasuda, A. Tsukazaki, R. Yoshimi, K. Kondou, K.S. Takahashi, Y. Otani, M. Kawasaki, Y. Tokura, Phys. Rev. Lett. **119**, 137204 (2017). https://doi.org/10.1103/PhysRevLett.119.137204. https://link.aps.org/doi/10.1103/PhysRevLett.119.137204
63. Y. Wang, D. Zhu, Y. Wu, Y. Yang, J. Yu, R. Ramaswamy, R. Mishra, S. Shi, M. Elyasi, K.L. Teo et al., Nat. Commun. **8**(1), 1364 (2017)
64. J. Han, A. Richardella, S.A. Siddiqui, J. Finley, N. Samarth, L. Liu, Phys. Rev. Lett. **119**(7), 077702 (2017)

65. D. Mahendra, R. Grassi, J.Y. Chen, M. Jamali, D.R. Hickey, D. Zhang, Z. Zhao, H. Li, P. Quarterman, Y. Lv et al., Nat. Mater. **17**(9), 800 (2018)
66. N.H.D. Khang, Y. Ueda, P.N. Hai, Nat. Mater. **17**(9), 808 (2018)
67. H. Wang, J. Kally, J.S. Lee, T. Liu, H. Chang, D.R. Hickey, K.A. Mkhoyan, M. Wu, A. Richardella, N. Samarth, Phys. Rev. Lett. **117**, 076601 (2016). https://doi.org/10.1103/PhysRevLett.117.076601. https://link.aps.org/doi/10.1103/PhysRevLett.117.076601
68. C. Tang, Q. Song, C.Z. Chang, Y. Xu, Y. Ohnuma, M. Matsuo, Y. Liu, W. Yuan, Y. Yao, J.S. Moodera, et al., Sci. Adv. **4**(6), eaas8660 (2018)
69. C.O. Avci, A. Quindeau, C.F. Pai, M. Mann, L. Caretta, A.S. Tang, M.C. Onbasli, C.A. Ross, G.S. Beach, Nat. Mater. **16**(3), 309 (2017)
70. Q. Shao, C. Tang, G. Yu, A. Navabi, H. Wu, C. He, J. Li, P. Upadhyaya, P. Zhang, S.A. Razavi, et al., Nat. Commun. **9** (2018)
71. P. Li, J. Kally, S. Zhang, T. Pillsbury, J. Ding, G. Csaba, J. Ding, J. Jiang, Y. Liu, R. Sinclair, C. Bi, A. DeMann, G. Rimal, W. Zhang, S. Field, J. Tang, W. Wang, O. Heinonen, V. Novosad, M. Wu, Sci. Adv. **5**, eaaw3415 (2019). https://doi.org/10.1126/sciadv.aaw3415
72. H. Wang, J. Kally, C. şahin T. Liu, W. Yanez, E.J. Kamp, A. Richardella, M. Wu, M.E. Flatté, N. Samarth, Phys. Rev. Res. **1**, 012014 (2019). https://doi.org/10.1103/PhysRevResearch.1.012014. https://link.aps.org/doi/10.1103/PhysRevResearch.1.012014
73. D. Hsieh, Y. Xia, D. Qian, L. Wray, J. Dil, F. Meier, J. Osterwalder, L. Patthey, J. Checkelsky, N.P. Ong et al., Nature **460**(7259), 1101 (2009)
74. K. Yasuda, R. Wakatsuki, T. Morimoto, R. Yoshimi, A. Tsukazaki, K. Takahashi, M. Ezawa, M. Kawasaki, N. Nagaosa, Y. Tokura, Nat. Phys. **12**(6), 555 (2016)
75. R. Wakatsuki, M. Ezawa, N. Nagaosa, Sci. Rep. **5**, 13638 (2015)

Chapter 5
Topological Phase Transitions Relevant to Quantum Anomalous Hall Effect

5.1 Introduction

The discovery and the subsequent studies of the QAH effect in magnetic TIs has established the exchange coupling between surface Dirac fermion states and magnetism and allows us to explore exotic functionalities arising from the surface-magnetized (massive) Dirac fermion states. In this chapter, we study quantum phase transitions relevant to the QAH effect and realize a new quantum phase based on magnetic TI heterostructures. Owing to the topological nature of bulk wavefunctions, 3D TIs possess the top and bottom surface degrees of freedom. When these surface states hybridize with each other through the quantum tunneling [1–4], the surface states open a gap. Such a hybridization-induced gap competes with the magnetic gap, leading to a quantum phase transition between a trivial insulator and a QAH insulator.

Next, we explore a new insulating quantum state termed an axion insulator which does not require the surface hybridization. The axion insulator state hosts a novel magnetoelectricity, so-called the topological magnetoelectric (TME) effect [5]. The TME effect is unique in that it originates from the orbital motion of the electrons due to the topological nature of crystalline solids as characterized by axion electrodynamics [6, 7] and also the response (susceptibility) is quantized to the half-integer quantum conductance of $e^2/2h$. The axion insulator state is realized when all the magnetization points outwards or inwards from the surface of a 3D TI [5, 8]. Such a special configuration of magnetization makes the TIs insulating because of no domain walls carrying the chiral edge states. To realize the axion insulator state, we employ heterostructure synthesis as investigated in Chap. 3. We introduce 'asymmetries' in the heterostructure and then realize the axion insulator upon the field-induced quantum phase transition from the QAH state. Such a quantum phase transition accompanies the gigantic magnetoresistance owing to the switching on/off of the chiral edge conduction.

© The Author(s), under exclusive license to Springer Nature Singapore Pte Ltd. 2022
M. Mogi, *Quantized Phenomena of Transport and Magneto-Optics in Magnetic Topological Insulator Heterostructures*, Springer Theses,
https://doi.org/10.1007/978-981-19-2137-7_5

5.2   Quantum Anomalous Hall Insulator–Trivial Insulator Transition via Magnetization Rotation

One way to realize the phase transition between the QAH insulator and the trivial insulator is controlling the sample thickness. When the TI films are thinned to below 5 QL, the hybridization gap dominates the magnetization gap [4]. Then the films become a trivial insulator.

An alternative way studied here is tuning the exchange gap upon magnetization rotation. Since the exchange gap Δ_{ex} originates from the perpendicular magnetic moments, tilting the magnetic moments would reduce Δ_{ex} as a function of the tilting angle. With tilting the angle, it makes Δ_{ex} smaller than the hybridization gap Δ_{hy}, then the sample would transit to the trivial insulator phase. We investigate this quantum phase transition by systematically changing the film thickness, clarifying the energy scale of Δ_{hy} against Δ_{ex}, namely Δ_{hy}/Δ_{ex}.

We fabricated magnetic TI heterostructures (Fig. 5.1a) with various total thicknesses $t = 7, 8, 10$, and 13 nm. While such heterostructures stabilize the QAH effect, the physics investigated here are not changed from that for uniformly doped samples. To rotate the magnetization, the direction of the magnetic field (B) is changed respect to the film plane (Fig. 5.1b). We applied $B = 2$ T so that the magnetization will follow the direction of the magnetic field. We define the angle θ away from the normal of the film plane.

We study the topological aspects of the phase transition by θ dependent Hall conductivity $\sigma_{xy}(\theta)$ (Fig. 5.1c), which measures the number of chiral edge channels associated with the Chern number C. When $\theta = 0°$, i.e. $\cos\theta = 1$, the QAH state emerges, where $\sigma_{xy} = e^2/h$ or $C = 1$. With tilting θ, σ_{xy} changes from e^2/h to 0, which is the quantum phase transition from the QAH insulator to the trivial insulator. A hallmark feature of this quantum phase transition is that σ_{xx} increases from zero at $\theta = 66.9°$, which is the manifestation of the gap closure of the topological phase transition. Another key point is that σ_{xy} at the transition angle is independent of temperature (T) (Fig. 5.1d), indicating a critical behavior of the transition. This transition resembles the QH plateau-plateau transitions driven by the changing of the Landau-level filing factor [10–13].

To clarify the connection between the present plateau-plateau transition and that for the QH state, we plot the data of Fig. 5.1c, d on the $\sigma_{xy} - \sigma_{xx}$ plane (Fig. 5.1e). On decreasing T, the flow of $(\sigma_{xy}(T), \sigma_{xx}(T))$ converges to either $(0, 0)$ or $(e^2/h, 0)$ with an unstable point at around $(0.5e^2/h, 0.5e^2/h)$. This behavior is similar to the features of the theoretically calculated renormalization group flow of conductivity tensor components in the QH system [14, 15].

Furthermore, σ_{xx} of both QAH ($\theta = 0.0°$) and trivial insulator ($\theta = 89.6°$) phases follow the Arrhenius-type T dependence as shown in Fig. 5.1f. Fitting the T dependence suggests the thermal activation energy defined as $\sigma_{xx} \propto \exp(E_a/k_B T)$ is $E_a = 55$ μeV and 26 μeV for $\theta = 0.0°$ and 89.6°, respectively. Figure 5.1g shows E_a at the intermediate angles. With increasing θ from 0°, i.e., decreasing $\cos\theta$ from 1, E_a linearly decreases at $\cos\theta > 0.4$ and then increases at $\cos\theta < 0.4$ as a function

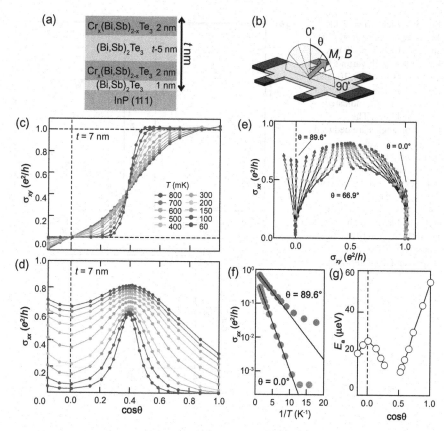

Fig. 5.1 **a** Schematic magnetic TI heterostructure. **b** Schematic alignments of magnetic field (B) and magnetization (M) with the angle θ from the direction normal to the sample. Field ($B = 2$ T) angle dependence of σ_{xy} **c** and σ_{xx} **d** for $t = 7$ nm sample plotted as a function of $\cos\theta$ under various temperatures from 60 to 800 mK. **e** Temperature driven conductivity ($\sigma_{xy}(T)$, $\sigma_{xx}(T)$) flows for various field angles (θ). **f** Log-scale plot of σ_{xx} as a function of $1/T$ for $\theta = 0.0°$ (pink) and 89.6° (blue). The black lines indicate the fitting results. (g) Variation of the activation energy E_a as a function of $\cos\theta$. Reprinted figure with permission from [9] Copyright 2018 by the American Physical Society

of $\cos\theta$. This behavior is consistent with the gap size tuning upon the magnetization rotation: $\Delta = \Delta_{\mathrm{ex}} - \Delta_{\mathrm{hy}} = \Delta^0_{\mathrm{ex}} \cos\theta - \Delta_{\mathrm{hy}}$, where Δ^0_{ex} is the original size of the exchange gap.

We then turn to the critical field angle θ_c, where the transition takes place as determined from the peak of σ_{xx}. When the transition is a result of the competition between the exchange gap and the hybridization gap, θ_c is given by $\Delta^0_{\mathrm{ex}} \cos\theta_c = \Delta_{\mathrm{hy}}$. Because Δ is smaller in thicker films, $\cos\theta_c$ is expected to decrease with increasing the film thickness t. We observed such a trend as shown in Fig. 5.2a, b, where θ_c is indicated by black triangles. We plot the $\cos\theta_{\mathrm{peak}}$ as a function of t (Fig. 5.2c). The solid curve in Fig. 5.2c is drawn using the extrapolation of the reported Δ_{hy}

Fig. 5.2 Field ($B = 2$ T) angle dependence of σ_{xy} (**a**) and σ_{xx} (**b**) at $T = 60$ mK for samples with various film thicknesses $t = 7, 8, 10$, and 13 nm. The data are offset for clarity. Zeros are indicated by dashed lines. The transition angles are deduced from the σ_{xx} peaks as highlighted by triangles in (**b**). **c** Film thickness dependence of the transition angles. The solid curve shows $\cos\theta_c = \Delta_{\text{hy}}(t)/\Delta_{\text{ex}}^0$. Reprinted figure with permission from [9] Copyright 2018 by the American Physical Society

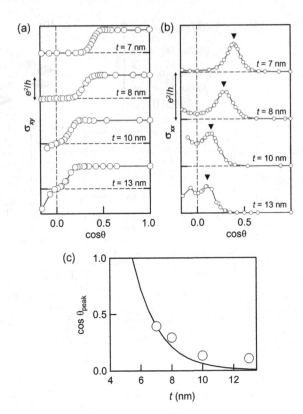

measured by ARPES in (Bi, Sb)$_2$Te$_3$ in the form of $\Delta_{\text{hy}} \propto \exp(-t/t_0)$ [4, 16] and the reported value of $\Delta_{\text{ex}}^0 = 30$ meV in (Bi, Sb)$_2$Te$_3$ doped with 10% Cr [17], which has the similar Cr concentration to our films. The data of the present study and the solid curve are roughly consistent. Furthermore, the value of σ_{xx} stays finite even at $\cos\theta = 0$ in the 10- and 13-nm-thick films. These observations thus conclude that the trivial insulator phase originates from the hybridization between the top and bottom surfaces.

5.3 Creation of Axion Insulator State

The axion insulator which may exhibit an exotic quantized topological magnetoelectric effect (TME effect) is one of the most interesting quantum phases [5, 18, 19]. A special configuration of the magnetization is required to realize the axion insulator state; the magnetizations should point inwards or outwards from the surface so that all the surfaces surrounding the 3D TI are gapped. Then, the 3D TI becomes an

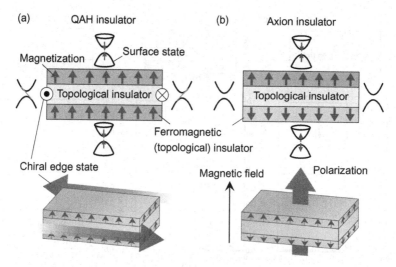

Fig. 5.3 Schematic illustrations of a quantum anomalous Hall (QAH) insulator (**a**) and an axion insulator (**b**) in magnetic TI thin films. In an axion insulator, a magnetic field induces electric polarization and an electric field induces magnetization as the TME effect

insulator, termed an axion insulator. However, experimental observation of the TME effect remains unexplored due to difficulties in realizing the axion insulator state.

In the case of a thin-film form, the above condition of inward or outward magnetization is achieved with the anti-parallel magnetization alignment of the top and bottom surfaces, as shown in Fig. 5.3b; namely magnetizations are to point up on the top surface and point down on the bottom surface. In addition to the gap opening at the top and bottom surfaces by the perpendicular magnetization, the side surface is can be gapped due to the quantum confinement effect. Thus, all the surface states are gapped to form an axion insulator with zero Hall conductivity (σ_{xy}) and longitudinal conductivity (σ_{xx}). Recent theories have shown that the TME effect can occur in such a thin film with a quantum confinement gap when the magnetizations are anti-parallel to each other. Thus, control of magnetism in the 3D magnetic TI thin films provides a key to the realization of the axion insulator state [20, 21].

To realize the anti-parallel magnetization on the top and bottom surfaces of the 3D magnetic TI, we engineered magnetic heterostructures of Cr-doped (Bi, Sb)$_2$Te$_3$ (BST) thin films. We doped Cr (12%) selectively to 2-nm-thick layers in the vicinity of the top and bottom surfaces, as shown in Fig. 5.4b, so that the surface electrons on the top (bottom) surface mainly and strongly interact with the upper (lower) Cr-doped layer through the exchange interaction. Such a modulation doping of magnetic moments enhances the stability of the QAH state (Fig. 5.3a) up to several kelvin. Here, as a new attempt, we introduce a vertical asymmetry to the heterostructure; the bottom Cr-doped BST layer is 1 nm away from the bottom surface, while the top Cr-doped BST layer is in contact with the top surface. The asymmetry leads to different coercive fields (B_c) in the two layers. Moreover, we make sure that the non-

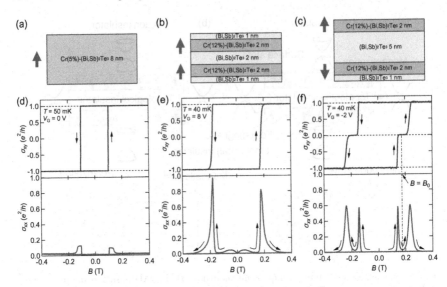

Fig. 5.4 a–c Schematic layouts of Cr-doped TI ((Bi, Sb)$_2$Te$_3$) thin films used for the experiments. The Cr density (%) and Bi:Sb ratio of respective films are 5% 22:78 (**a**) and 12%, 26:74 (**b, c**). **d–f** Magnetic field (B) dependence of Hall conductivity (σ_{xy}) and longitudinal conductivity (σ_{xx}) in the respective thin films shown in (**a–c**) at the lowest temperature (40–50 mK) of a dilution refrigerator. The applied gate voltage is fixed at the charge neutrality of the respective thin films ($V_G = 0$ V for (**a**), 8 V for (**b**), −2 V for (**c**)). The measured range of B is between 2 T and −2 T for magnetization training. B taking a minimal value of σ_{xx} in the ZHP state is defined as $B = B_0$. Reprinted from Springer Nature [22], Copyright 2017 Macmillan Publishers Limited, part of Springer Nature

magnetic BST separation layer (5 nm) between the Cr-doped BST layers has enough thickness to reduce the interlayer magnetic coupling (see Fig. 5.5). The total film thickness of 10 nm can also weaken the coupling between the top and bottom surface states. For the control experiments, we prepared a vertically symmetric magnetic TI heterostructure film with a thinner BST separation layer (2 nm) (Fig. 5.4b) and an 8-nm-thick uniformly Cr(5%)-doped (Bi$_{0.22}$Sb$_{0.78}$)$_2$Te$_3$ single-layer film (Fig. 5.4a).

In terms of magnetotransport measurements, we can verify the realization of the QAH state and an axion insulator depending on the magnetization direction of two magnetic layers in the magnetic TI heterostructure. When the applied perpendicular magnetic field (B) is large enough to align the magnetizations of the two magnetic layers parallel, then the QAH state with $\sigma_{xy} = e^2/h$ is expected to appear. With decreasing magnetic field close to the coercive field B_c, the magnetization of one of the Cr-doped layers is reversed to form the anti-parallel magnetization alignment, as shown in Fig. 5.3b, which can host the axion insulator with $\sigma_{xy} = 0$. With further decreasing in magnetic field, the magnetization of the other Cr-doped layer is also reversed and the magnetizations become parallel again, returning to the QAH state. Therefore, by monitoring the B dependence of σ_{xy}, the axion insulator state can be observed. For the transport measurement, we prepared Hall-bar and Corbino-disk

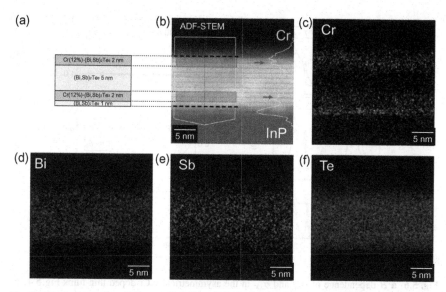

Fig. 5.5 **a** Schematic layout of a magnetic TI heterostructure studied here. Dotted lines are the guide to the eyes for the ADF-STEM image shown in (**b**). **b** ADF-STEM image of a magnetic TI heterostructure shown in a and a line profile of Cr distribution. Red shaded regions are the layers where we intended to dope Cr. The red arrows indicate the peaks of the line profile of Cr. **c–f**, Distribution maps of each element, Cr (**c**), Bi (**d**), Sb (**e**) and Te (**f**) studied by EDX. Reprinted from Springer Nature [22], Copyright 2017 Macmillan Publishers Limited, part of Springer Nature

devices with electrostatic gates so as to tune the chemical potential in the magnetization gap.

Let us begin with the results of the homogeneously Cr-doped 8-nm-thick film (Fig. 5.4a). The QAH state appears with σ_{xy} quantized to e^2/h and almost vanishing σ_{xx} at $B = 0$ T (Fig. 5.4d). σ_{xy} sharply transits between $C = e^2/h$ and $-e^2/h$, accompanied by a single peak in σ_{xx}. This result indicates that the magnetization reversal occurs at once at the single coercive field B_c. Similarly, in the symmetric magnetic TI heterostructure (Fig. 5.4b), quantized σ_{xy} reverses sharply accompanied by a single σ_{xx} peak at around $B = B_c$ (Fig. 5.4e), indicating that the magnetization reversal of the two magnetic layers occurs simultaneously at a single B_c. We infer that the two magnetic layers happen to have the same B_c and/or that an interlayer magnetic coupling between them remains more or less because of the relatively thin BST separation layer (2 nm).

In contrast to the above results, in the asymmetric magnetic TI heterostructure (Fig. 5.4c), we observe clear zero Hall conductivity ($\sigma_{xy} = 0$) plateaus (ZHPs) between the $\sigma_{xy} = e^2/h$ plateaus of the QAH state (Fig. 5.4f). Corresponding to the double transitions of σ_{xy} (e^2/h to 0 and 0 to $-e^2/h$), σ_{xx} exhibits two peaks and becomes almost zero between them. These behaviors of σ_{xx} and σ_{xy} directly indicate the formation of the axion insulator state. Although the Cr densities of the two magnetic TI layers are the same, we anticipate that the vertical asymmetry of Cr

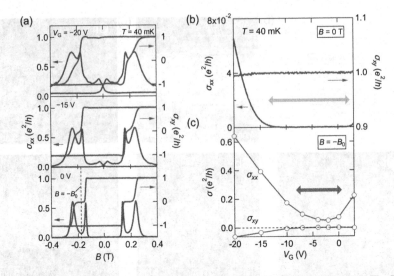

Fig. 5.6 a B dependence of σ_{xy} and σ_{xx} in the asymmetrically Cr-doped thin films Fig. 5.4c at various gate voltages, $V_G = 0$, -15 and -20 V at $T = 40$ mK. **b, c** Gate voltage (V_G) dependence for the QAH state at $B = 0$ T after training of magnetization by applying $B = 2$ T (**b**) and for the ZHP state at $B = -B_0$ (**c**) plotted from the B scan data of σ_{xy}. Double-headed arrows in **b** (green) and **c** (purple) are guides to the eyes for the V_G width sustaining the QAH and ZHP states, respectively. Reprinted from Springer Nature [22], Copyright 2017 Macmillan Publishers Limited, part of Springer Nature

doping brings about the difference in the magnetic anisotropy between the separated magnetic layers. In the following, we focus on the ZHP state in this asymmetric heterostructure with the thicker separation layer (5 nm) (Fig. 5.4c).

To clarify the relation between the two quantized states, the ZHP and QAH states, we investigated their dependences on the chemical potential by applying a gate voltage (V_G). Figure 5.6a shows the B dependence of σ_{xy} and σ_{xx} at $V_G = 0$, -15 and -20 V. In negative V_G, σ_{xx} in the QAH state gradually deviates from zero and the width of σ_{xy} in ZHPs becomes narrower. We show the V_G dependence of σ_{xy} and σ_{xx} under the QAH state ($B = 0$ T) and the ZHP state ($B = B_0$) in Fig. 5.6b, c, respectively. Close to the charge neutrality point, judged from the QAH state, σ_{xx} also takes a minimum value and $\sigma_{xy} = 0$ in the ZHP state. Thus, both the QAH state and the ZHP state are stabilized near the charge neutrality point.

We further investigated the ZHPs in the devices with alternative geometries. Two-terminal conductance in the Hall-bar (Fig. 5.7a) and the Corbino-disk (Fig. 5.7b) were measured by applying an a.c. voltage (10 μV, 3 Hz). Figure 5.7c shows the two-terminal conductance (G_{2T}) of the Hall-bar device used in Fig. 5.6. In the QAH state, the value of G_{2T} ($= 0.95e^2/h$) is close to e^2/h due to the presence of the chiral edge states (a slight deviation can be attributed to an additional contact resistance). In the ZHP state around $B = B_0$ ($= -0.18$ T), on the other hand, G_{2T} becomes zero (to be exact, $G_{2T} < 0.0002e^2/h$), indicating the absence of the chiral edge channels connecting the two electrodes owing to the quantum confinement effect

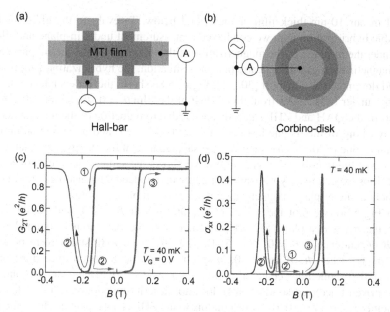

Fig. 5.7 a, b Schematic illustration of the two-terminal conductivity measurement in a Hall-bar geometry (**a**) and a Corbino-disk geometry (**b**) by applying a 10-μV a.c. voltage. **c, d** Two-terminal conductivity (G_{2T}) (**c**) and σ_{xx} (**d**) as a function of B of the minor loop (from red line to blue line) and the full loop from $B = 2$ T to -2 T (from red line to gray line) at $T = 40$ mK without applying a gate voltage. The arrows and the numbers are guides to the eyes indicating the direction and sequence of the field scan. The value of σ_{xx} is calculated by the relation: $\sigma_{xx} = 2\pi\sigma_0\ln(r_{out}/r_{in})$, where σ_0 represents a measured conductance and $r_{out}(r_{in})$ represents the outer (inner) radius of the Corbino-disk. Reprinted from Springer Nature [22], Copyright 2017 Macmillan Publishers Limited, part of Springer Nature

on the side surface [23–26]. Furthermore, we conducted a minor loop measurement starting from $B = 2$ T and reversing the magnetic field from $B = B_0$ (stable ZHP state) to a positive B. We found that the ZHP state with almost zero conductance is maintained even at $B = 0$ T. This highly stable insulation of the ZHP state is advantageous for detection of the TME effect. Figure 5.7d shows results of minor and major loop measurements on the Corbino-disk which directly provides σ_{xx} without the calculation of the tensor relation needed in the four-terminal measurement. The outline of the loops is consistent with the results of the four-terminal measurement in the Hall-bar, implying that almost zero conductivity is realized.

Similar transitions from the QAH states to a ZHP state have been reported in the studies using uniformly Cr-doped ultrathin ($\sim$5, 6 QL) TI films [27, 28] where strong hybridization between top and bottom surface states is anticipated. The transition is interpreted as a topological phase transition of the surface state to a trivial insulator phase where the magnetization gap becomes smaller than hybridization gap during the magnetic multi-domain state around B_c [27–29]. Accordingly, the hybridization gap is essential for the occurrence of the ZHPs in their interpretation. However, this cannot be applied to our observation because the hybridization gap is negligibly small in our 10-nm-thick film, as described below. To evaluate the effect of the possible hybridization gap, we conducted a measurement under in-plane and tilted B. Under the in-plane B, the magnetization of the magnetic TI also lies in plane and the magnetization gap disappears. In such a situation, the hybridization gap, if any, would dominate the transport [30, 31]. Figure 5.8a shows the B dependence of σ_{xx} and σ_{xy} under field angles from the normal of the film $\theta = 0, 79, 85$ and $90°$. With tilting B, the QAH and ZHP states disappear due to reduction of the magnetization gap by tilting of the magnetization. At $\theta = 90°$, $\sigma_{xy} = 0$ and σ_{xx} takes an almost constant value ($\sim 0.5e^2/h$), suggesting disappearance of the energy gap - that is, an absence of the hybridization gap. Figure 5.9 shows the same experiment for 8-nm-thick films expecting a hybridization effect, which indeed demonstrates the smaller σ_{xx} both at the in-plane magnetization and zero-Hall plateau states.

In Fig. 5.8b, we plot the values of σ_{xy} and σ_{xx} at $B = 2$ T with variation of θ on the $\sigma_{xy} - \sigma_{xx}$ plane [32–34], together with the data points obtained from the B dependence at $\theta = 0°$ (Fig. 5.4f). The QAH and ZHP states correspond to $(\sigma_{xy}, \sigma_{xx}) = (\pm e^2/h, 0)$ and $(0, 0)$, respectively. In the B-driven trace (blue open circles), two semicircles centered at $(\sigma_{xy}, \sigma_{xx}) = (\pm e^2/2h, 0)$ appear. By contrast, the θ-driven trace (red solid circles) does not describe the double semicircles and the point $(\sigma_{xy}, \sigma_{xx}) = (0, 0)$ corresponding to the ZHP is not reached. This indicates that the hybridization gap is too small to establish the ZHP state of the hybridization origin. On the other hand, in the symmetric structures, only a single semi-circle is observed due to the degeneracy of the top and bottom surface states (no formation of the axion insulator state) as shown in Fig. 5.10.

To discuss more quantitatively, in Fig. 5.8c we show the temperature dependence of σ_{xx} in the Corbino-disk under the QAH state at $B = 0$ T (green solid triangles), the ZHP states at $B = -B_0$ (black open circles) and 0 T (blue open circles) from B dependence, and an in-plane ($\theta = 90°$) magnetic field at $B = 2$ T (red solid squares). Under the in-plane B, where no magnetization gap is expected, σ_{xx} decreases slightly with decreasing temperature, but tends to saturate at around $0.5e^2/h$. This result suggests that the hybridization gap in the magnetic TI heterostructure studied here is small compared to the measurement temperature ($T = 40$ mK). By contrast, in the QAH and ZHP states, σ_{xx} decreases dramatically and becomes zero ($\sigma_{xx} < 0.0002e^2/h$) with lowering T. The dependence in the QAH and ZHP states can be ascribed to thermal activated conduction with the respective energy scales of 1.52 K in the QAH state ($B = 0$ T), 0.45 K ($B = -B_0$), and 0.61K ($B = 0$ T) in the ZHP states. It should be noted that the ZHP state at $B = 0$ T is more stable than the ZHP state at $B = -B_0$. The observed enhanced stability of the ZHP states at $B = 0$ T may be accounted for by the magnetic stability of the magnetic TI heterostructure. The magnitude of the energy gap in the ZHP state is proportional to the perpendicular component of the magnetic moments. At $B = -B_0$ which is close to one of the two (top- and bottom-layer) coercive fields, the magnetic moments may partly tilt from the original direction, resulting in the reduction of the energy gap on average. Hence, by removing the external magnetic field at $B = 0$ T, the anti-parallel magnetic configuration becomes more stable, as reflected in the slightly higher activation energy.

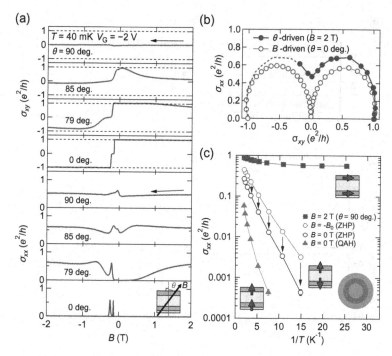

Fig. 5.8 **a** B (swept from 2 T to −2 T) dependence of σ_{xy} and σ_{xx} when under field angles $\theta = 0°, 79°, 85°, 90°$ at $T = 40$ mK and $V_G = -2$ V in the Hall-bar geometry. θ is defined as the angle off the normal of the film. **b** Comparison between $(\sigma_{xy}, \sigma_{xx})$-plots by the magnetic-field-driven results at $\theta = 0°$ (out-of-plane) (blue open circles) and the θ-driven (from 0 to 96°) (red solid circles) results at $B = 2$ T. The θ-driven curve is symmetrized and connected by a dashed line from the results between 0 and 96°. **c** Temperature ($T = 530, 470, 380, 300, 200, 130, 90, 70$, and 40 mK) dependence of σ_{xx} under the QAH and ZHP states and under an in-plane magnetic field ($B = 2$ T) on a logarithmic scale as a function of $1/T$. Red solid squares show the data of σ_{xx} under an in-plane magnetic field. Blue and black open circles show σ_{xx} under ZHP state at $B = -B_0$ and 0 T respectively. Green solid triangles show σ_{xx} under QAH state at $B = 0$ T. The data are taken in the Corbino-disk geometry (as in Fig. 5.7c). On lowering T, σ_{xx} in the ZHP state (~ 40 mK) and the QAH state (~ 100 mK) becomes smaller than $0.0002e^2/h$. Reprinted from Springer Nature [22], Copyright 2017 Macmillan Publishers Limited, part of Springer Nature

5.4 Quantum Anomalous Hall Insulator–Axion Insulator Transition with Gigantic Magnetoresistance

Next, we realize a robust axion insulator state by tailoring a tricolor structure of magnetic TI thin film as shown in Fig. 5.11a. Such a stabilized axion insulator state leads to gigantic magnetoresistance upon the field induced phase transition from the QAH state. The tricolor structure was designed so that the two separated magnetic layers have a large difference in B_c. Here, we used Cr- and V-doped BST, which were reported to exhibit the QAH effect in uniformly doped films [35, 36]. The values of B_c

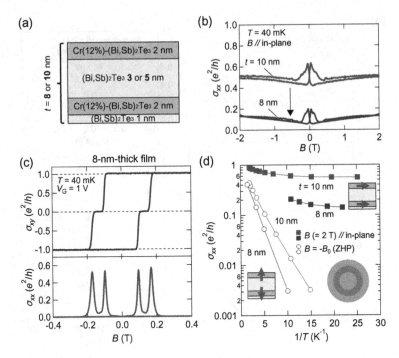

Fig. 5.9 **a** Schematic layout of magnetic TI heterostructures studied in (**b**), (**c**) and (**d**). **b** In-plane B dependence of the 8- (blue) and 10- (red) nm-thick films measured at $T = 40$ mK. **c** Perpendicular B dependence of σ_{xy} and σ_{xx} in a 8-nm-thick film. **d** Temperature (T) dependence of σ_{xx} for the two samples measured in Corbino-disks under in-plane magnetic field ($B = 2$ T) and under ZHP states ($B = -B_0$) on a logarithmic scale as a function of $1/T$. Solid squares show the data of 8- (blue) and 10- (red) nm-thick films under in-plane magnetic field. Open circles show the data of 8- (blue) and 10- (red) nm-thick films under ZHP state. Reprinted from Springer Nature [22], Copyright 2017 Macmillan Publishers Limited, part of Springer Nature

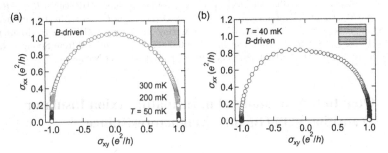

Fig. 5.10 **a**, **b** (σ_{xy}, σ_{xx})-plots from the results of magnetic field dependent σ_{xy} and σ_{xx} of the uniformly Cr-doped film (**a**) shown in Fig. 5.4d and of the symmetric magnetic TI heterostructure (**b**) shown in Fig. 5.4e. Reprinted from Springer Nature [22], Copyright 2017 Macmillan Publishers Limited, part of Springer Nature

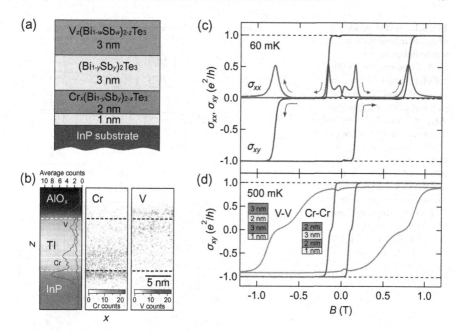

Fig. 5.11 **a** Schematic structure of MBE-grown TI (BST) film doped with Cr and V used in the present study. The compositions: $(x, y, z, w) = (0.24, 0.74, 0.15, 0.80)$. **b** Cross-sectional STEM image (left) and net-count mappings of Cr (center) and V (right) in the Cr-V-doped tricolor BST film obtained by EDX. The averages along x of the net counts of Cr and V are indicated by green and red, respectively. **c** B dependence of σ_{xy} and σ_{xx} of the Cr-V-doped BST film at 60 mK. **d** B dependence of σ_{xy} at 500 mK in the Cr-Cr-doped [$Cr_{0.24}(Bi_{0.26}Sb_{0.74})_{1.76}Te_3$] (shown in green) and the V-V-doped [$V_{0.15}(Bi_{0.20}Sb_{0.80})_{1.85}Te_3$] (shown in red) bicolor TI heterostructures. Their growth profiles are schematically shown as insets with the respective thicknesses of the layers. Reprinted from [37] by The Author(s) licensed under CC BY 4.0

have been reported as 0.2 and 1 T for the Cr- and V-doped BST, respectively. To ensure the magnetic decoupling of the two layers, we insert a nonmagnetic BST between the Cr- and V-doped layers. In this tricolor structure (Fig. 5.11a), the electrons on the bottom surface state interact with the Cr ions and those on top surface state interact with V ions.

We grew the tricolor-structure film of Cr-doped BST (2 nm)/BST (3 nm)/V-doped BST (3 nm) (Cr-V-doped BST) on the InP(111) substrate (Fig. 5.11a). To align the charge neutrality points for the top and bottom surface states, we modulated the Bi/Sb ratio at the Cr- and the V-doped layer. A 1-nm-thick BST was grown as a buffer layer between the substrate and the film for the improvement of the quality, providing an enhancement in the observable temperature of the QAH effect. We also prepared Cr-Cr- and V-V-doped bicolor TI heterostructure films (insets of Fig. 5.11d) for comparison. Cross-sectional STEM and EDX observations prove that Cr and V ions are separately distributed in the film as expected (Fig. 5.11b).

Fig. 5.12 a B dependence of σ_{xy} at representative temperatures in the Cr-V-doped tricolor TI film. **b** Color mapping of σ_{xy} for the Cr-V-doped TI tricolor film and plots of peaks for $d\sigma_{xy}/dB$, which correspond to coercive fields (defined as B_{c1} for smaller and B_{c2} for larger) of the Cr-V-doped tricolor film (shown in red) and the Cr-Cr-doped bicolor film (shown in green) in the T-B plane. Reprinted from [37] by The Author(s) licensed under CC BY 4.0

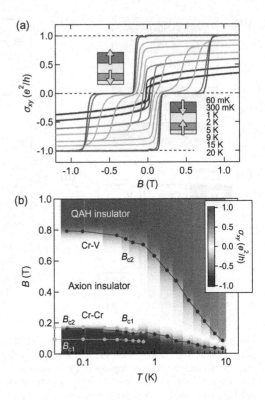

Figure 5.11c shows the B dependence of σ_{xy} and σ_{xx} in a Hall bar device. Double-step transitions are observed in the B dependence of σ_{xy}. σ_{xx} has double peaks at the B of σ_{xy} transitions; otherwise, σ_{xx} is almost zero. Between the two transitions in σ_{xy}, the ZHP is observed in a range of $0.19\,\text{T} < B < 0.72\,\text{T}$. Figure 5.11d shows the transport properties of the V-V-doped (3 nm) and Cr-Cr-doped (2 nm) BST bicolor heterostructures at 500 mK. The B_c of the Cr-Cr-doped and V-V-doped bicolor films are 0.2 and 0.8 T, respectively. The transition fields of the Cr-V-doped film (Fig. 5.11c) nearly correspond to the B_c of the Cr-Cr-doped ($\sim 0.2\,\text{T}$) and V-V-doped ($\sim 0.8\,\text{T}$) bicolor films. The correspondence unambiguously shows that the ZHP originates from the antiparallel magnetization alignments of the Cr- and V-doped layers. As shown in Fig. 5.11d, precursors of the ZHP are seen in the B dependence of σ_{xy} in the Cr-Cr-doped and V-V-doped films as well. However, the antiparallel magnetization in these films is probably induced by the unintentional difference in B_c between the upper and lower magnetically doped layers; therefore, the observed B range for the ZHP is much narrower that for the Cr-V-doped tricolor film.

In Fig. 5.12a, we show the B dependence of σ_{xy} at various temperatures from 60 mK to 20 K. A kink structure is observed around the sign changes of σ_{xy} below 9 K, which develops into the ZHP with decreasing temperature. Below 300 mK, the ZHP is well established and the quantization of σ_{xy} to $\pm e^2/h$ is seen at $B = 0$ T. In

Fig. 5.12b, the values of B_c are plotted in the $T - B$ plane for the Cr-V-doped and the Cr-Cr-doped bicolor films. The B_c is determined from the peak positions of $d\sigma_{xy}/dB$. The smaller and the larger B_c values are defined as B_{c1} and B_{c2}, respectively. The area corresponding to the ZHP between the QAH states is highlighted as a white region. Compared to the Cr-Cr-doped bicolor film, the area of the ZHP is far extended in the T-B plane as designed. The B_{c1} of the Cr-V-doped tricolor film roughly coincides with the $B_{c1,2}$ of the Cr-Cr-doped bicolor film. The area of the ZHP is extended not only along the magnetic field axis but also along the temperature axis in the Cr-V-doped film. Improved stability of the antiparallel magnetization configuration is probably responsible for the increase in the observable temperature of the ZHPs.

We then investigate the robustness of the ZHP states against the application of electric field using two-terminal transport measurements on the Corbino disk. In Fig. 5.13, we show current-voltage (I-V) characteristics of the Cr-V-doped tricolor film at 60 mK. The I-V curve exhibits an insulating behavior up to $|V| \sim 10$ mV. The application of $|V| > 10$ mV breaks down the insulating state, resulting in a steep increase in I. We compare the I-V characteristics of the Cr-V-doped tricolor film with those observed in the Cr-Cr-doped bicolor film. The breakdown voltage of the Cr-V-doped film is about five times larger than that of the Cr-Cr-doped bicolor film ($|V| \sim 2$ mV). Focusing on the I-V curves in the low voltage regime (Fig. 5.13, insets), the two-terminal resistance reaches about 300 MΩ' in the Cr-V-doped tricolor film, which is about six times larger than that of the Cr-Cr-doped tricolor film (50 MΩ). Therefore, the stability of the axion insulator state against the electric field is also improved in the Cr-V-doped tricolor film compared to the Cr-Cr-doped bicolor film. Although the relevant electric field direction for the TME effect is perpendicular to the film, the improved robustness against the in-plane electric field strongly suggests that the obtained axion insulator state is perhaps robust also to the perpendicular electric fields.

The large two-terminal resistance (R_{2T}) of the axion insulator state leads to a gigantic magnetoresistance (MR) ratio. Figure 5.14a shows the B dependence of the R_{2T} of the Hall bar device (width $W = 300$ mm and length $L = 1$ mm) at various temperatures. The R_{2T} exhibits a butterfly shape when B is swept. The low resistance value of $R_{2T} = 28$ kΩ under the parallel magnetization configuration jumps up by about 9000 times to 260 MΩ under the antiparallel magnetization configuration. When the field scan is stopped at 0.38 T, the resistance value keeps increasing up to 2.8 GΩ with the elapsed time, perhaps because of the sample temperature, which was increased during the B scan, cooling down; this gives the MR ratio [= $(R_{2T}^{AP} - R_{2T}^{P})/R_{2T}^{P}$, where AP and P denote the antiparallel and parallel magnetization configurations, respectively] exceeding 10^5 (or 10^7%). The gigantic MR is demonstrated within 2 T, being distinct from the extremely large MR (XMR) reported for other topological materials, such as WTe$_2$ and NbP, by applying a high magnetic field above 50 T [38, 39]. Even at 0.5 K, the MR ratio remains as high as 1800%, which is comparable to the XMR at $B = 2$ T. One other important and distinctive feature of the MR ratio in the present system is that the low resistance value R_{2T}^{P} is mainly dominated by the Hall resistance. Therefore, R_{2T}^{P} takes a value close to h/e^2 and does not strongly depend on the dimensions of the Hall bar. On the other hand, the high resistance value R_{2T}^{AP} scales with the aspect ratio of the Hall bar as usual. As a result, the MR ratio associated with the QAH/axion insulator transition, which corresponds to the switching on/off of the chiral edge channels, becomes dependent on the Hall bar dimensions. The axion insulator state with such a high resistance is maintained even when the field is swept back to $B = 0$ T. Figure 5.14b shows the temperature (T) dependence of R_{2T} at $B = 0$ T under the antiparallel and parallel magnetization configurations (T dependence of ρ_{xx} and ρ_{yx} measured in the Hall bar are plotted together). The resistance under antiparallel magnetization is highly sensitive to T. By increasing T from 60 up to 400 mK, R_{2T}^{AP} decreases by three orders of magnitude. The high resistance values in Fig. 5.14a measured by scanning B probably suffered from the temperature increase during the B scan. This behavior contrasts to the R_{2T} under the parallel configuration, which is mainly dominated by the Hall resistance and is only weakly dependent on T.

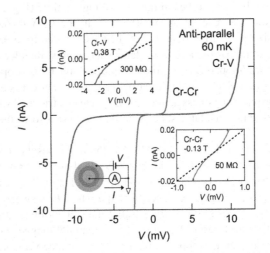

Fig. 5.13 *I-V* curves under the antiparallel magnetization configuration of the Cr-V-doped tricolor BST (red) and the Cr-Cr-doped bicolor BST (green) heterostructures, taken in a Corbino disk device at $T \sim 50$ mK. For stabilization of the antiparallel magnetization configuration, the external magnetic field is fixed at the minimum of conductivity (for the Cr-V-doped BST, $B = -0.38$ T; for the Cr-Cr-doped BST, $B = -0.13$ T). Upper (lower) inset: Expanded scale of the *I-V* curve of the Cr-V-doped (Cr-Cr-doped) BST heterostructure. Reprinted from [37] by The Author(s) licensed under CC BY 4.0

Fig. 5.14 a B dependence of R_{2T} between current terminals in the Hall bar of the Cr-V-doped BST film measured at various temperatures ($T = 60, 140, 200, 300,$ and 500 mK). **b** T dependence of R_{2T} in the parallel/antiparallel magnetization configurations under zero magnetic field. ρ_{yx} and ρ_{xx} measured by a four-terminal probe are also shown. Inset: minor loop hysteresis of σ_{xy} at $T = 60$ mK, reversing B at -0.38 T. Reprinted from [37] by The Author(s) licensed under CC BY 4.0

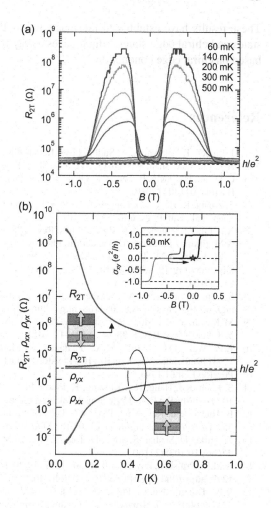

5.5 Conclusion

We have investigated the quantum phase transition of the QAH insulator phase to a trivial insulator phase and to an axion insulator phase. When the film sample has about 7 nm in thickness, the QAH phase can transit to the trivial insulator phase upon magnetization rotation. Thickness dependence of the phase transition has clarified that the trivial insulator phase originates from the surface hybridization gap. Furthermore, through the fabrication of magnetic heterostructure, we have realized the axion insulator by tuning the magnetization direction of the surface magnetic layers. The transition between the QAH phase and the axion insulator phase accompanies the gigantic magnetoresistance due to the switching on/off of chiral edge conduction.

These results have established the magnetic control of the surface Dirac fermion states and chiral edge states, which paves a way for design novel electrical circuits based on chiral edge conduction.

References

1. J. Linder, T. Yokoyama, A. Sudbø. Phys. Rev. B **80**(20), 205401 (2009)
2. H.Z. Lu, W.Y. Shan, W. Yao, Q. Niu, S.Q. Shen, Phys. Rev. B **81**(11), 115407 (2010)
3. C.X. Liu, H. Zhang, B. Yan, X.L. Qi, T. Frauenheim, X. Dai, Z. Fang, S.C. Zhang, Phys. Rev. B **81**(4), 041307 (2010)
4. Y. Zhang, K. He, C.Z. Chang, C.L. Song, L.L. Wang, X. Chen, J.F. Jia, Z. Fang, X. Dai, W.Y. Shan et al., Nat. Phys. **6**(8), 584 (2010)
5. X.L. Qi, T.L. Hughes, S.C. Zhang, Phys. Rev. B **78**(19), 195424 (2008)
6. A.M. Essin, J.E. Moore, D. Vanderbilt, Phys. Rev. Lett. **102**(14), 146805 (2009)
7. A.M. Essin, A.M. Turner, J.E. Moore, D. Vanderbilt, Phys. Rev. B **81**, 205104 (2010). https://doi.org/10.1103/PhysRevB.81.205104. https://link.aps.org/doi/10.1103/PhysRevB.81.205104
8. S. Coh, D. Vanderbilt, A. Malashevich, I. Souza, Phys. Rev. B **83**, 085108 (2011).https://doi.org/10.1103/PhysRevB.83.085108. https://link.aps.org/doi/10.1103/PhysRevB.83.085108
9. M. Kawamura, M. Mogi, R. Yoshimi, A. Tsukazaki, Y. Kozuka, K.S. Takahashi, M. Kawasaki, Y. Tokura, Phys. Rev. B **98**(14), 140404 (2018)
10. H.P. Wei, D.C. Tsui, M.A. Paalanen, A.M.M. Pruisken, Phys. Rev. Lett. **61**, 1294 (1988). https://doi.org/10.1103/PhysRevLett.61.1294. https://link.aps.org/doi/10.1103/PhysRevLett.61.1294
11. S. Koch, R.J. Haug, K.v. Klitzing, K. Ploog, Phys. Rev. B **46**, 1596 (1992). https://doi.org/10.1103/PhysRevB.46.1596. https://link.aps.org/doi/10.1103/PhysRevB.46.1596
12. B. Huckestein, Rev. Mod. Phys. **67**, 357 (1995). https://doi.org/10.1103/RevModPhys.67.357. https://link.aps.org/doi/10.1103/RevModPhys.67.357
13. M. Hilke, D. Shahar, S. Song, D. Tsui, Y. Xie, D. Monroe, Nature **395**, 675 (1998). https://doi.org/10.1038/27160
14. A.M.M. Pruisken, Phys. Rev. Lett. **61**, 1297 (1988). https://doi.org/10.1103/PhysRevLett.61.1297. https://link.aps.org/doi/10.1103/PhysRevLett.61.1297
15. B.P. Dolan, Nucl. Phys. B **554**(3), 487 (1999). https://doi.org/10.1016/S0550-3213(99)00326-0. http://www.sciencedirect.com/science/article/pii/S0550321399003260
16. F. Zhang, C.L. Kane, E.J. Mele, Phys. Rev. B **86**, 081303 (2012). https://doi.org/10.1103/PhysRevB.86.081303. https://link.aps.org/doi/10.1103/PhysRevB.86.081303
17. I. Lee, C.K. Kim, J. Lee, S.J. Billinge, R. Zhong, J.A. Schneeloch, T. Liu, T. Valla, J.M. Tranquada, G. Gu et al., Proc. Natl. Acad. Sci. USA **112**(5), 1316 (2015)
18. X. Wan, A.M. Turner, A. Vishwanath, S.Y. Savrasov, Phys. Rev. B **83**, 205101 (2011). https://doi.org/10.1103/PhysRevB.83.205101. https://link.aps.org/doi/10.1103/PhysRevB.83.205101
19. X. Wan, A. Vishwanath, S.Y. Savrasov, Phys. Rev. Lett. **108**, 146601 (2012). https://doi.org/10.1103/PhysRevLett.108.146601. https://link.aps.org/doi/10.1103/PhysRevLett.108.146601
20. T. Morimoto, A. Furusaki, N. Nagaosa, Phys. Rev. B **92**, 085113 (2015). https://doi.org/10.1103/PhysRevB.92.085113. https://link.aps.org/doi/10.1103/PhysRevB.92.085113
21. J. Wang, B. Lian, X.L. Qi, S.C. Zhang, Phys. Rev. B **92**, 081107 (2015).https://doi.org/10.1103/PhysRevB.92.081107.
22. M. Mogi, M. Kawamura, R. Yoshimi, A. Tsukazaki, Y. Kozuka, N. Shirakawa, K. Takahashi, M. Kawasaki, Y. Tokura, Nat. Mater. **16**(5), 516 (2017)
23. H.H. Fu, J.T. Lü, J.H. Gao, Phys. Rev. B **89**, 205431 (2014). https://doi.org/10.1103/PhysRevB.89.205431. https://link.aps.org/doi/10.1103/PhysRevB.89.205431

24. T. Morimoto, A. Furusaki, N. Nagaosa, Phys. Rev. Lett. **114**, 146803 (2015). https://doi.org/10.1103/PhysRevLett.114.146803. https://link.aps.org/doi/10.1103/PhysRevLett.114.146803

25. J. Wang, B. Lian, H. Zhang, S.C. Zhang, Phys. Rev. Lett. **111**, 086803 (2013). https://doi.org/10.1103/PhysRevLett.111.086803. https://link.aps.org/doi/10.1103/PhysRevLett.111.086803

26. X. Kou, S.T. Guo, Y. Fan, L. Pan, M. Lang, Y. Jiang, Q. Shao, T. Nie, K. Murata, J. Tang, Y. Wang, L. He, T.K. Lee, W.L. Lee, K.L. Wang, Phys. Rev. Lett. **113**, 137201 (2014). https://doi.org/10.1103/PhysRevLett.113.137201. https://link.aps.org/doi/10.1103/PhysRevLett.113.137201

27. Y. Feng, X. Feng, Y. Ou, J. Wang, C. Liu, L. Zhang, D. Zhao, G. Jiang, S.C. Zhang, K. He, X. Ma, Q.K. Xue, Y. Wang, Phys. Rev. Lett. **115**, 126801 (2015). https://doi.org/10.1103/PhysRevLett.115.126801. https://link.aps.org/doi/10.1103/PhysRevLett.115.126801

28. X. Kou, L. Pan, J. Wang, Y. Fan, E. Choi, W.L. Lee, T. Nie, K. Murata, Q. Shao, S.C. Zhang, K. Wang, Nat. Commun. **6**, 8474 (2015). https://doi.org/10.1038/ncomms9474

29. J. Wang, B. Lian, S.C. Zhang, Phys. Rev. B **89**, 085106 (2014). https://doi.org/10.1103/PhysRevB.89.085106. https://link.aps.org/doi/10.1103/PhysRevB.89.085106

30. A. Kandala, A. Richardella, S. Kempinger, C.X. Liu, N. Samarth, Nat. Commun. **6**, 7434 (2015). https://doi.org/10.1038/ncomms8434

31. X. Feng, Y. Feng, J. Wang, Y. Ou, Z. Hao, C. Liu, Z. Zhang, L. Zhang, C. Lin, J. Liao, Y. Li, l. Jun, S.H. Ji, X. Chen, X. Ma, S.C. Zhang, Y. Wang, K. He, Q.K. Xue, Adv. Mater. **28**, 6386 (2016). https://doi.org/10.1002/adma.201600919

32. K. Nomura, N. Nagaosa, Phys. Rev. Lett. **106**, 166802 (2011). https://doi.org/10.1103/PhysRevLett.106.166802. https://link.aps.org/doi/10.1103/PhysRevLett.106.166802

33. S. Kivelson, D.H. Lee, S.C. Zhang, Phys. Rev. B **46**, 2223 (1992). https://doi.org/10.1103/PhysRevB.46.2223. https://link.aps.org/doi/10.1103/PhysRevB.46.2223

34. C.P. Burgess, R. Dib, B.P. Dolan, Phys. Rev. B **62**, 15359 (2000). https://doi.org/10.1103/PhysRevB.62.15359. https://link.aps.org/doi/10.1103/PhysRevB.62.15359

35. C.Z. Chang, J. Zhang, X. Feng, J. Shen, Z. Zhang, M. Guo, K. Li, Y. Ou, P. Wei, L.L. Wang et al., Science **340**(6129), 167 (2013)

36. C.Z. Chang, P. Tang, X. Feng, K. Li, X.C. Ma, W. Duan, K. He, Q.K. Xue, Phys. Rev. Lett. **115**, 136801 (2015). https://doi.org/10.1103/PhysRevLett.115.136801. https://link.aps.org/doi/10.1103/PhysRevLett.115.136801

37. M. Mogi, M. Kawamura, A. Tsukazaki, R. Yoshimi, K.S. Takahashi, M. Kawasaki, Y. Tokura, Sci. Adv. **3**(10), eaao1669 (2017)

38. M. Ali, J. Xiong, S. Flynn, J. Tao, Q. Gibson, L. Schoop, T. Liang, N. Haldolaarachchige, M. Hirschberger, N. Ong, R. Cava, Nature **514**, 205 (2014). https://doi.org/10.1038/nature13763

39. C. Shekhar, A. Nayak, Y. Sun, M. Schmidt, M. Nicklas, I. Leermakers, U. Zeitler, W. Schnelle, J. Grin, C. Felser, B. Yan, Nat. Phys. **11**, 645 (2015). https://doi.org/10.1038/nphys3372

Chapter 6
Half-Integer Quantized Electrodynamics in 3D Topological Insulator

6.1 Introduction

A 3D TI supports a single gapless Dirac fermion at each surface as a consequence of the nontrivial Z_2 topological nature of wavefunctions in the insulating bulk. Unlike the 2D lattice systems where paired Dirac cones exist at different points of momentum space, in the 3D TI, paired Dirac cones appear on the opposite surfaces, specifically top and bottom surfaces in a crystal of thin-film form. The QAH and axion insulator states as studied in Chap. 5 are understood by such a Dirac fermions picture, where the Hall conductance is given by a half-integer quantized value per a single massive Dirac cone. In this picture, the QAH and axion insulator states are understood as the sum and the subtraction of the half-quantized Hall conductivity, respectively. However, the half-integer quantization of the single Dirac cone itself has been elusive, because the two Dirac cones on top and bottom surfaces contribute to the measured values in ordinary experiments. Nevertheless, if we can measure only a single surface state, the half-integer QHE is expected to be measured. In this chapter, we demonstrate that the half-integer quantization of Hall conductance associated with the parity anomaly can be realized in 3D TIs. Furthermore, we study the bulk-edge correspondence for the half-integer quantized state via terahertz (THz) magneto-optical and transport measurements.

One of the experimental methods to observe the half-integer quantization is the THz magneto-optical Faraday and Kerr rotation measurements. The magneto-optical polarization rotations integrate the rotatory contributions. When the light passes through the surface states with the quantized Hall conductivity, the polarization rotation angles contain the quantized value [1–3].

© The Author(s), under exclusive license to Springer Nature Singapore Pte Ltd. 2022
M. Mogi, *Quantized Phenomena of Transport and Magneto-Optics in Magnetic
Topological Insulator Heterostructures*, Springer Theses,
https://doi.org/10.1007/978-981-19-2137-7_6

6.2   Quantized Faraday and Kerr Rotations in Quantum Anomalous Hall State

First, we study the magneto-optical Faraday and Kerr rotations in the QAH state for the magnetic TI thin films by using time-domain THz spectroscopy. In the previous study, we used the penta-layer of Cr-doped $(Bi, Sb)_2Te_3$ thin film (Fig. 3.1c) [4]. In this study, we used the tetra-layer of Cr-doped $(Bi, Sb)_2Te_3$ thin film as shown in Fig. 5.4c, where we have much improved a signal-to-noise ratio compared to the previous study.

With the use of the low-energy limit of light that does not excite the carriers owing to the sizable magnetization gap, the magneto-optical polarization rotation angles are expected to be quantized. Theoretical investigations for the magneto-optical effect in the QAH state have shown that the expected Faraday and Kerr rotation angles are varied with the experimental geometries, such as a semi-infinite TI bulk [5], a free-standing TI film [2], and a TI film on a substrate [3]. The derivation uses the boundary conditions for electromagnetic waves, where the contribution of half-quantized Hall current generated by electric fields of the light is included as presented in Chap. 1. The expected rotation angles for the various geometries are summarized in Fig. 6.1.

A practical geometry for the time-domain THz spectroscopy of the Faraday and Kerr rotation measurements is the use of multiple reflections inside the substrate (Fig. 6.1d) because the transmission of a THz pulse can be divided into two pulses, of which the first pulse is rotated by only the Faraday effect while the second pulse reflected at the backside of the substrate is rotated by both the Faraday and Kerr effects. As a result, the Faraday and Kerr rotations are given by,

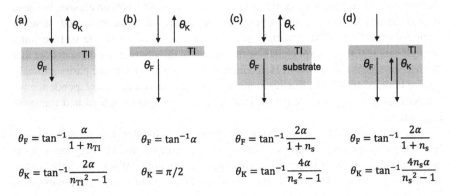

$$\theta_F = \tan^{-1}\frac{\alpha}{1+n_{TI}} \qquad \theta_F = \tan^{-1}\alpha \qquad \theta_F = \tan^{-1}\frac{2\alpha}{1+n_s} \qquad \theta_F = \tan^{-1}\frac{2\alpha}{1+n_s}$$

$$\theta_K = \tan^{-1}\frac{2\alpha}{n_{TI}{}^2-1} \qquad \theta_K = \pi/2 \qquad \theta_K = \tan^{-1}\frac{4\alpha}{n_s{}^2-1} \qquad \theta_K = \tan^{-1}\frac{4n_s\alpha}{n_s{}^2-1}$$

Fig. 6.1 a–d Quantized Faraday and Kerr rotation angles calculated for various QAH sample geometries of the semi-infinite TI (**a**), the free-standing TI film (**b**), the TI film on the substrate (**c**, **d**). For (**c**, **d**), Kerr rotation measurements conducted by the reflection (**c**) and the transmission (**d**) configurations. n_{TI} and n_s denote the refractive indices of the TI and the substrate, respectively

$$\theta_F = \tan^{-1} \frac{2\alpha}{1 + n_s}, \tag{6.1}$$

$$\theta_K = \tan^{-1} \frac{4n_s\alpha}{n_s^2 - 1}, \tag{6.2}$$

where α is the fine-structure constant ($\sim 1/137$) and n_s is the refractive index of the substrate. While these two quantities depend on the materials parameter n_s of the substrate, we can extract α by combining them [3],

$$\alpha = \frac{\tan\theta_F \tan\theta_K - \tan^2\theta_F}{\tan\theta_K - 2\tan\theta_F}. \tag{6.3}$$

We now present the Faraday and Kerr rotation spectra for the magnetic TI heterostructure exhibiting the QAH effect (the inset of Fig. 6.2); the spectra are shown in Fig. 6.2a, b as a function of the incident light energy ($\hbar\omega$, where $\hbar = h/2\pi$) taken at zero magnetic field ($\mu_0 H = 0$ T) and temperature $T = 1$ K once after training the magnetization up to 2 T, which was obtained by analyzing the time-domain waveforms of the transmitted THz pulses (see Chap. 2). We find that the real parts of Faraday θ_F and Kerr θ_K rotation angles with the negligible imaginary parts (ellipticities) of η_F and η_K are consistent with the theoretical values of the quantized Faraday and Kerr rotations: $\tan^{-1}\left(\frac{\alpha}{n_s+1}\right) = 3.26$ mrad and $\tan^{-1}\left(\frac{4n_s\alpha}{n_s^2-1}\right) = 9.21$ mrad, respectively, where $n_s \sim 3.46$ is the refractive index of InP substrates. By applying magnetic fields up to 1 T, the rotation angles do not change from those at $\mu_0 H = 0$

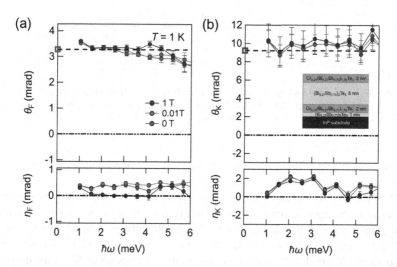

Fig. 6.2 **a, b** Complex Faraday ($\theta_F + i\eta_F$) **a** and Kerr ($\theta_F + i\eta_F$) **b** rotation spectra for a magnetic Cr modulation doped TI film exhibiting the QAH effect, as schematically shown in the inset of (**b**), under external magnetic fields with a slight variation ($\mu_0 H = 0, 0.01,$ and 1 T). The error bars represent standard error of mean

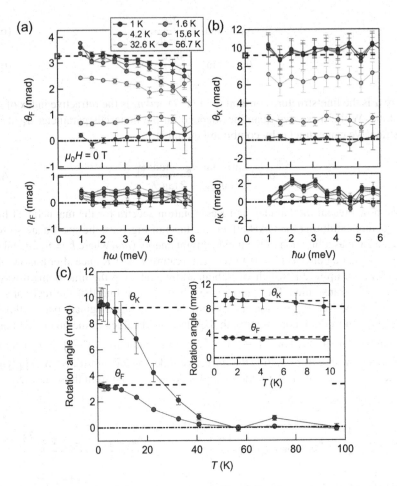

Fig. 6.3 **a, b** θ_F, η (**a**) and θ_F, η (**b**) spectra at $\mu_0 H = 0$ T and at various temperatures ($T = 1$, 1.6, 4, 2, 15.6, 32.6, and 56.7 K for the QAH insulator. **c** Temperature dependence of θ_F and θ_K taken at $\hbar\omega = 2$ meV. The error bars represent standard error of mean

T, being consistent with the QAH effect emerged even at zero magnetic field. These results are in accord with the above prediction.

We then investigate the temperature dependence of the magneto-optical rotation angles at zero magnetic field. As shown in Fig. 6.3a, b, with increasing the temperature, the rotation angles decrease from the quantization values. At 56.7 K, both θ_F and θ_K become zero. In Fig. 6.3c, we show the variation of θ_F and θ_K as a function of temperature. We find that the zero rotation angles correspond to T_C for the magnetic TI film (~ 50 K) and that the quantization subsists up to about 4.2 K.

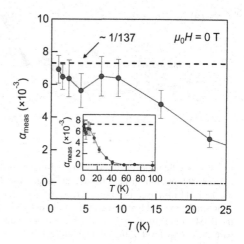

Fig. 6.4 Measured fine-structure constant α_{meas} which is calculated by using the relation of $\alpha_{meas} = (\tan \theta_F \tan \theta_K - \tan^2 \theta_F)/(\tan \theta_K - 2 \tan \theta_F)$ from the data shown in Fig. 6.3c

Lastly, we convert θ_F and θ_K to α by using the relation of Eq. 6.3. As shown in Fig. 6.4, the measured fine-structure constant α_{meas} roughly takes the expected value of α below the quantization temperature.

6.3   Half-Integer Quantization of Hall Conductance Associated with the Parity Anomaly

To explore the half-quantized magneto-optical effect, we now study a modified magnetically-doped TI termed 'semi-magnetic' TI: By doping magnetic ions only in the vicinity of the top surface (Fig. 6.5a), only the top-surface Dirac cone is gapped out while keeping the bottom-surface Dirac cone gapless. Since only one of the paired Dirac cones is gapped, this semi-magnetic TI can be an arena for the half-integer quantization in magneto-optical effects.

In the semi-magnetic TI heterostructure films consisting of $(Bi, Sb)_2 Te_3$ and Cr-doped $(Bi, Sb)_2 Te_3$ (Fig. 6.5b), the magnetic element Cr was modulation-doped only near the top surface (2 nm) [6]. The exchange interaction between the surface electrons and the magnetic ions (Cr) opens an energy gap on the top surface Dirac cone. The Fermi energy E_F was carefully tuned so that E_F lies within the magnetic gap of the top surface Dirac cone. The energy level of the Dirac point for the bottom surface (E_{WP}) remains tunable by changing the Bi:Sb ratio (x).

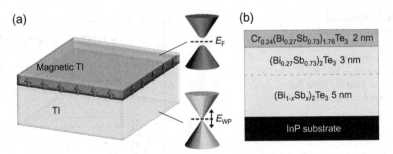

Fig. 6.5 **a** Schematic illustration of a semi-magnetic TI, where the top surface Dirac state is gapped, whereas the bottom surface Dirac state is gapless. **b** Schematic layout of a MBE-grown semi-magnetic TI film used for the experiments. The total thickness of 10 nm makes the hybridization between the top and bottom surface states negligible. In the upper part (5 nm from the top surface), the Bi:Sb ratio is fixed so that E_F is located in the magnetically-induced gap, whereas, in the lower part, the energy level of the Dirac point (E_{WP}) can be tuned by varying Sb fraction, x

6.3.1 Half-Quantized Faraday and Kerr Rotations

Figure 6.6c displays the Faraday rotation angle θ_F. We observe that θ_F takes around 1.6 mrad, being almost independent of the incident light frequency. This value is almost half of that for the QAH system as shown in Fig. 6.6a, b [4] and integer ($\nu = 1$) QH systems [7]. It also agrees with the prediction of the half-quantized Faraday rotation angle: The Faraday rotation, for small rotation angles, is described by the summation of the dynamical 2D Hall conductivity $\sigma_{xy}^{2D}(\omega, z)$ along the light path (z-direction), $\tan \theta_F = \frac{1}{(n_s+1)c\varepsilon_0} \sum_i \sigma_{xy}^{2D}(\omega, z_i)$ (z_i : z coordinate of the i-th layer), where ε_0 is the permittivity of the vacuum and c is the speed of light. Given that the top and bottom surface states are well decoupled and that the gapped top surface state has half quantized Hall conductivity while the gapless bottom surface and bulk states have no contribution, the rotation angle gives rise to $\theta_F = \tan^{-1}\left(\frac{\alpha}{n_s+1}\right) \approx 1.60$ mrad ($\alpha = \frac{e^2}{2\varepsilon_0 hc} \approx \frac{1}{137}$ is the fine structure constant), which is consistent with the experimentally measured values around 1.6 mrad.

Similarly, the Kerr rotation angle θ_K was measured by utilizing multiple reflections of the THz pulses inside the substrate (see Chap. 2). As shown in Fig. 6.6d, the value of θ_K is around 4.4 mrad, which is also almost half of that for the QAH state (Fig. 6.6b). This also agrees with the predicted half-integer quantization for a thin film limit ($\theta_K = \tan^{-1}\left(\frac{2n_s\alpha}{n_s^2-1}\right) \approx 4.45$ mrad). The ellipticities η_F and η_K are quite small compared with θ_F and θ_K, respectively, suggesting no resonance feature due to a sufficient magnetic gap opening of the top surface state compared to the incident THz light energy ($\hbar\omega < 5$ meV). Thus, the observation of the half quantized rotations under $\mu_0 H = 0$ T is consistent with the picture that only the gapped top surface works as a source of magneto-optical rotations and the contribution of the gapless bottom surface to the rotations is negligibly small (see Appendix 6.5).

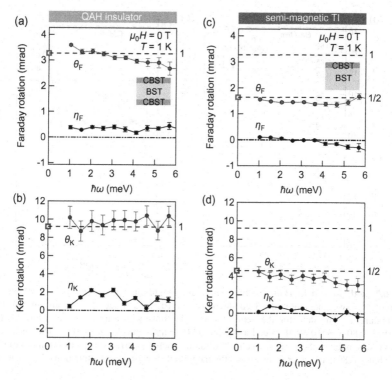

Fig. 6.6 a, b Complex Faraday ($\theta_F + i\eta_F$) (**a**) and Kerr ($\theta_K + i\eta_K$) (**b**) rotation spectra for a semi-magnetic TI film ($x = 0.89$) at $T = 1$ K and at $\mu_0 H = 0$ T. The real (θ_F, θ_K) and imaginary (η_F, η_L) parts represent the polarization rotation angle and the ellipticity of the transmitted light, respectively. The dashed lines and open squares indicate the expected half-quantized rotation angles ($\theta_F = \tan^{-1}\frac{\alpha}{n_s+1} = 1.60$ mrad and $\theta_K = \tan^{-1}\frac{2n_s\alpha}{n_s^2-1} = 4.45$ mrad)

6.3.2 Half-Integer Quantization in Transport

The almost frequency-independent magneto-optical rotation spectra motivated us to study the Hall conductance at a zero-frequency limit, namely dc electrical transport measurements. In the magneto-optical rotation measurements, the observed Faraday and Kerr rotations can be regarded as a consequence of the topological nature in the interior of the gapped top surface. In dc electrical transport measurements, on the other hand, only the gapless bottom surface state where the electrical current is running through is to be probed. Thus, confirming the correspondence between the magneto-optical rotation and dc electrical transport measurements is important for understanding the surface-bulk correspondence of the half-integer quantized electrodynamics in TI.

In Fig. 6.7a, b, we show the temperature (T) dependence of the Hall (ρ_{yx}) and longitudinal (ρ_{xx}) resistivities at $\mu_0 H = 0$ T, respectively, for five samples with

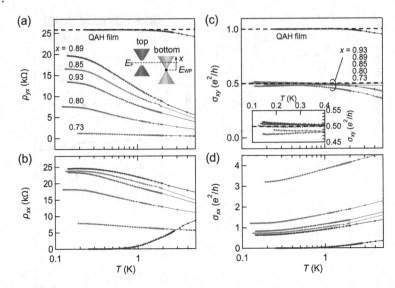

Fig. 6.7 a–d, Zero-field Hall resistivity ρ_{yx} (**a**), longitudinal resistivity ρ_{xx} (**b**), Hall conductivity σ_{xy} (**c**), and longitudinal conductivity σ_{xx} (**d**) as a function of logarithmic scale of temperature T. The data for the semi-magnetic TI films with various values of Bi:Sb ratio $x = 0.73, 0.80, 0.85, 0.89, 0.93$ and a typical QAH sample (purple) are shown. The inset of (**a**) shows a schematic of the semi-magnetic TI film, in which the carrier type is electron (hole) when $x \leq 0.85 (\geq 0.89)$. The inset of (**c**) shows the magnified view of σ_{xy} as a function of a linear scale of T

different Bi:Sb ratios (x) corresponding to different E_{WP} (energy level of the Dirac point for the bottom surface state); see the inset of Figs. 6.7a and 6.5b. Apparently, there are no characteristic features common to the respective samples. By contrast, when resistivity is converted to conductivity, the values of σ_{xy} for all the samples converge to the half quantum conductance $0.5e^2/h$ at low temperatures as clearly seen in Fig. 6.7c. The quantization occurs below 2 K, which is comparable to the typical quantization temperature of the QAH effect (purple curves in Figs. 6.7a, b). This result indicates that regardless of different E_{WP}, σ_{xy} is quantized to the half-integer value as long as E_F resides in the magnetic gap of the top surface state. Figure 6.7d shows the data of longitudinal conductivity σ_{xx} which shows a relatively large variation with x. σ_{xx} can be attributed to the dissipative electron transport in the bottom surface where the carrier densities are changed by x.

6.3.3 Comparison Between Transport and Magneto-Optics

In Fig. 6.8a, we directly compare the results of the magneto-optical rotation and the dc electric transport measurements using an identical semi-magnetic TI sample

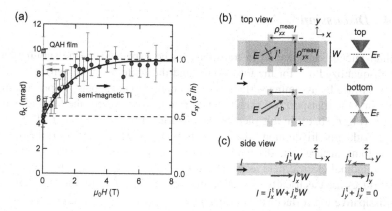

Fig. 6.8 a Magnetic field $\mu_0 H$ dependence of σ_{xy} (black line) measured by dc transport and θ_K at $\hbar\omega = 2$ meV (blue circles) at $T = 1$ K. **b, c** Schematics of the top view (**b**) and the side view (**c**) of the current flow in the parallel conduction model. The magnetic (top) and non-magnetic (bottom) layers are drawn in green and blue colors, respectively

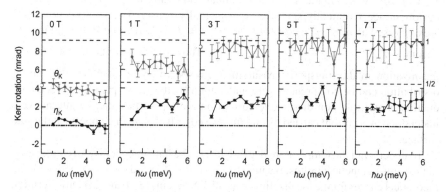

Fig. 6.9 Spectra of θ_K and η_K under various magnetic fields ($0.5\,\text{T} \leq \mu_0 H \leq 7\,\text{T}$)

($x = 0.93$). The vertical axis is scaled using the relation $\frac{\sigma_{xy}}{\tan\theta_K} = c\varepsilon_0 \frac{n_s^2-1}{2n_s}$. The results of magneto-optics and transport show a good agreement with each other, indicating that the half-integer quantization is verified regardless of measurement methods. This observation ensures that the bulk-surface correspondence is maintained in the semi-magnetic TI. When perpendicular magnetic fields are applied (see also Fig. 6.9), both σ_{xy} and θ_K are almost doubled and exhibit the integer quantization: $\sigma_{xy} = e^2/h$ and $\theta_K = \tan^{-1}\left(\frac{4n_s\alpha}{n_s^2-1}\right) = 9.20$ mrad. The doubled rotations under strong magnetic fields can be understood by the Landau level formation on the bottom surface which works as an additional source of the half-quantized rotations.

6.3.4 Discussion

The half-integer quantization in transport can be explained by the summation of the half-quantized σ_{xy} for the top surface and the finite σ_{xx} with zero σ_{xy} for the bottom surface by assuming surface parallel conduction channels. To provide more insight into the transport, we sketch the current distribution for the top and bottom surfaces (Fig. 6.8b, c). In this picture, the electric current $I = (j_x^t + j_x^b)W$ (W: the sample width) is driven along the x-direction (Fig. 6.9d), where t (b) denotes the top (bottom) surface. The dissipative bottom surface current density j_x^b generates the longitudinal electric field as described by $E_x = j_x^b/\sigma_{xx}^b$. Since the top and bottom surfaces are electrically shorted at the side surface, this E_x also generates the non-dissipative top surface Hall current $j_y^t = \sigma_{yx}^t E_x = -\frac{e^2}{2h} E_x$ because the top surface state is gapped. Importantly, because the y-direction current densities on the top and bottom surfaces must compensate with each other, $j_y^t + j_y^b = 0$ (Fig. 6.9d). This additional dissipative current density j_y^b generates the transverse electric field $E_y = j_y^b/\sigma_{xx}^b$ which induces the 'longitudinal' Hall current $j_x^t = \sigma_{xy}^t E_y = \frac{e^2}{2h} E_y$ on the top surface. Eventually, since the experimentally measured resistivity components are expressed as $\rho_{xx}^{\text{meas}} = E_x/(I/W)$ and $\rho_{yx}^{\text{meas}} = E_y/(I/W)$, the Hall conductivity becomes $\sigma_{xy}^{\text{meas}} = \frac{\rho_{yx}^{\text{meas}}}{(\rho_{xx}^{\text{meas}})^2 + (\rho_{yx}^{\text{meas}})^2} = \sigma_{xy}^t = \frac{e^2}{2h}$. Thus, $\sigma_{xy}^{\text{meas}}$ exhibits the half-integer quantization in the transport measurement irrespective of σ_{xx}^b. While the current distribution can be comprehensively understood by the above parallel surface conduction model, the non-dissipative current on the top surface can be alternatively understood as a current carried by an effective edge-like state that appears at the side surface.

To understand the temperature dependence of the half-integer quantized transport in the semi-magnetic TI films, we map $(\sigma_{xy}(T), \sigma_{xx}(T))$ and compare them with the theoretical curves for the localization in usual 2D systems [8, 9] and 2D Dirac fermion systems [10] with broken time-reversal symmetries as shown in Fig. 6.10a and b, respectively. Whereas $(\sigma_{xy}(T), \sigma_{xx}(T))$ for the QAH film follows the theoretical curves for the usual 2D system (Fig. 6.10a) as reported earlier [11], those for the semi-magnetic TI films converge to $\sigma_{xy} = 0.5e^2/h$. In addition, with even further decreasing the temperature, σ_{xy} show signatures to increase from $0.5e^2/h$. These behaviors contradict the scaling trajectory of the usual 2D system [8, 9]. Instead, this can be viewed as the sum of the scaling behaviors for the two 2D Dirac fermions with magnetic impurities [10], where σ_{xx} approaches zero and half-quantized σ_{xy} becomes a stable fixed point for each of the Dirac fermion (Fig. 6.10b). Since the effect of the magnetic impurities on the bottom surface is weak, the increase of its contribution to σ_{xy} is small even at the lowest temperature (blue arrow line in Fig. 6.10a); this is the reason why the plateau of σ_{xy} is observed. Furthermore, if the effect of magnetic impurities on the bottom surface state is completely absent, σ_{xx} is expected to increase as the temperature is lowered. Nonetheless, σ_{xx} decreases, and as the temperature is even much lowered σ_{xy} tends to further increase slightly across this half-quantized value. Thus, these features indicate that the system is the almost

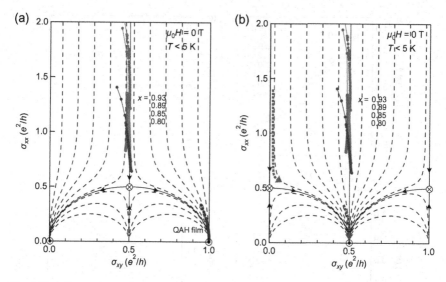

Fig. 6.10 **a** Temperature-driven conductivity flows of $(\sigma_{xy}(T), \sigma_{xx}(T))$ for the semi-magnetic TI films ($x = 0.80$, 0.85, 0.89, and 0.93) and the QAH film. Broken lines show the renormalization group flow for conventional 2D systems [9]. **b** The same plots for the semi-magnetic TI films. The broken lines are the renormalization group flow for 2D Dirac fermion systems with broken time-reversal symmetry [10]. Red and blue broken arrows are illustration of flows describing the behaviors for the top and bottom surface Dirac fermions in the semi-magnetic TI, respectively

decoupled two Dirac fermions with large (top surface) and small (bottom surface) σ_{xy} with broken time-reversal symmetry.

Finally, we mention the accuracy of the half quantized σ_{xy}. Experimentally, the measured deviation of σ_{xy} from $e^2/2h$ was 2.6% (the inset of Fig. 6.7c). Because σ_{xx} remains finite, the accuracy of the half-integer quantization cannot be so high as the QAH effect, in principle. Furthermore, in the ground state, namely $T \rightarrow 0$ K, all the electronic states would be localized as seen in the decrease of σ_{xx} with lowering temperature (Fig. 6.7d) due to broken time-reversal symmetry as a whole system [10], such as a tiny amount of magnetic (Cr) impurities possibly contained in the bottom surface, or else due to a tiny energy gap forms on the bottom surface by top-bottom surface hybridization (see Appendix 6.5). This would cause a shift of σ_{xy} to either e^2/h or 0, respectively. Nevertheless, the experimentally observed half-integer quantization has a surprisingly high stability in the parameter ranges of the present experiments by varying E_F (Fig. 6.7), T (down to 50 mK), and the sample size (W = 10 μm to 1.5 mm) (Fig. 6.11).

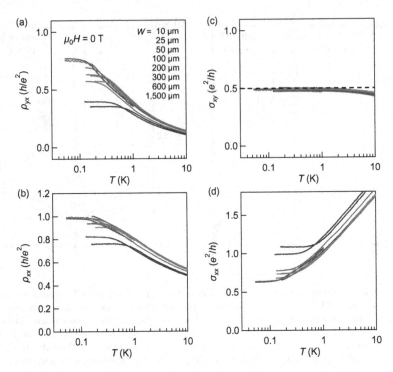

Fig. 6.11 **a–d** ρ_{yx} (**a**), ρ_{xx} (**b**), σ_{xy} (**c**), and σ_{xx} (**d**) as a function of T in a logarithmic scale at $\mu_0 H = 0$ T in semi-magnetic TI films ($x = 0.85$) with various sample width W

6.4   Conclusion

We have experimentally demonstrated the parity anomaly with the half-integer quantization of Hall conductance in the semi-magnetic TI. The two different methods of bulk-sensitive magneto-optical spectroscopy and surface-sensitive transport measurements ensure the bulk-surface correspondence of the half-integer quantization phenomenon. The semi-magnetic TI structure with the parity anomaly can be a platform for exotic single Dirac fermion physics, such as dyon particles arising from the fractional topological magneto-electric effect [12] and more accessible non-Abelian Majorana edge states [13–16].

6.5 Appendix

6.5.1 Derivation of Low-Energy Faraday and Kerr Rotations in Semi-magnetic Topological Insulator

We derive the low-energy limit of Faraday and Kerr rotation angles for a model of the semi-magnetic TI on a substrate as shown in Fig. 6.12. By using boundary conditions for the electric and magnetic fields traveling across the film and the substrate, we construct a transfer matrix of the light polarizations [17].

We first consider that the light propagates along z-direction and is transmitted through the interfaces (x-y plane) with electrical conductivities σ_{xx} and σ_{xy} (Fig. 6.12). We define the transfer matrix (T^I) connecting between the electric fields at the left-hand side (L) and those at the right-hand side (R) of the interface:

$$\begin{pmatrix} E_x^{+R} \\ E_y^{+R} \\ E_x^{-R} \\ E_y^{-R} \end{pmatrix} = T^I \begin{pmatrix} E_x^{+L} \\ E_y^{+L} \\ E_x^{-L} \\ E_y^{-L} \end{pmatrix}, \tag{6.4}$$

where $+(-)$ denotes the direction of the light propagation, $+z(-z)$. TI is derived by using the boundary conditions for the electric and magnetic fields following the Maxwell equations: $\nabla \times \boldsymbol{E} = -\partial \boldsymbol{B}/\partial t$, and $\nabla \times \boldsymbol{H} = \boldsymbol{j} + \partial \boldsymbol{D}/\partial t = \sigma \boldsymbol{E}\delta(z - z_0) + \varepsilon\partial \boldsymbol{E}/\partial t$, where ε is a dielectric constant of the medium, z_0 is the position of

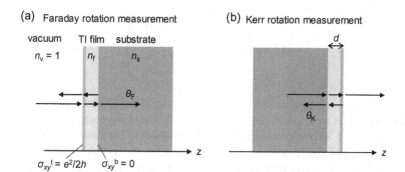

(a) Faraday rotation measurement   (b) Kerr rotation measurement

Fig. 6.12 **a** Schematic geometry for the Faraday rotation measurement in a TI/substrate structure. The light propagates along z-axis from the vacuum of which a refractive index $n_v = 1$. The reflection occurs at the interfaces between the TI film and the vacuum ($\sigma_{xy}^t = 1/2(e^2/h)$ and $\sigma_{xx}^t = 0$) and between the TI film and the substrate ($\sigma_{xy}^b = 0$ and $\sigma_{xx}^b \neq 0$). The transmitted light experiences the Faraday rotation. **b** Schematic geometry for the Kerr rotation measurement where the light comes from the substrate side. The light reflects at the top and bottom surfaces of the film, experiencing the Kerr rotation without a time delay between them because the film thickness d which we consider is far shorter than the wavelength of light

the interface, and σ is the 2×2 electrical conductivity tensor at the interface. Then, the boundary conditions are given by

$$E^L - E^R = 0, \tag{6.5}$$

$$\hat{z} \times \left(H^L - H^R\right) = \sigma E^L = \sigma E^R, \tag{6.6}$$

where $\hat{z}$ is the unit vector along the $+z$, c is the speed of light, ω is the frequency of light, n is the refractive index of the medium, and $E^{L(R)}$ and $H^{L(R)}$ are the left (right) hand side of electric and magnetic fields, respectively. By approximating the relative permeability constant in each region to 1, the electric and magnetic fields are given by $E^{\pm L(R)} = E_0^{\pm L(R)} e^{i(\pm k_z z + \omega t)}$ and $B^{\pm L(R)}(= \mu_0 H^{\pm L(R)}) = \pm \frac{n}{c} \hat{z} \times E^{\pm L(R)}$, where $k_z = n\omega/c$ is the wavenumber of light. By using Eqs. 6.4, 6.5, 6.6, $T^I(n_L, n_R, \sigma_{xx}, \sigma_{xy})$ ($n_{L(R)}$ is the refractive index of the left- (right-) hand side of the medium and σ_{xx}, σ_{xy} are conductivities in a unit of e^2/h) is described by

$$T^I = \frac{1}{2} \begin{pmatrix} 1 + \frac{n_L}{n_R} - \frac{2\alpha\sigma_{xx}}{n_R} & -\frac{2\alpha\sigma_{xy}}{n_R} & 1 - \frac{n_L}{n_R} - \frac{2\alpha\sigma_{xx}}{n_R} & -\frac{2\alpha\sigma_{xy}}{n_R} \\ \frac{2\alpha\sigma_{xy}}{n_R} & 1 + \frac{n_L}{n_R} - \frac{2\alpha\sigma_{xx}}{n_R} & \frac{2\alpha\sigma_{xy}}{n_R} & 1 - \frac{n_L}{n_R} - \frac{2\alpha\sigma_{xx}}{n_R} \\ 1 - \frac{n_L}{n_R} + \frac{2\alpha\sigma_{xx}}{n_R} & \frac{2\alpha\sigma_{xy}}{n_R} & 1 + \frac{n_L}{n_R} + \frac{2\alpha\sigma_{xx}}{n_R} & \frac{2\alpha\sigma_{xy}}{n_R} \\ -\frac{2\alpha\sigma_{xy}}{n_R} & 1 - \frac{n_L}{n_R} + \frac{2\alpha\sigma_{xx}}{n_R} & -\frac{2\alpha\sigma_{xy}}{n_R} & 1 + \frac{n_L}{n_R} + \frac{2\alpha\sigma_{xx}}{n_R} \end{pmatrix} \tag{6.7}$$

where α is the fine structure constant ($\alpha = e^2/2\epsilon_0 hc$). Then, considering the interference of the light due to multi-reflections inside the film, the transfer matrix T^B for the propagation in the TI bulk with thickness d is given by

$$T^B = \begin{pmatrix} e^{ik_z d} & 0 & 0 & 0 \\ 0 & e^{ik_z d} & 0 & 0 \\ 0 & 0 & e^{-ik_z d} & 0 \\ 0 & 0 & 0 & e^{-ik_z d} \end{pmatrix} \tag{6.8}$$

where k_z is the wavenumber of the light propagating to the TI film, d is the thickness of the film. Since we consider the long-wavelength (low-frequency) limit ($k_z d \to 0$), $T^B = I$, where I is the identity matrix. In the situation of the Faraday rotation measurement for a TI thin film on a substrate (Fig. 6.12a), the incident light is polarized along x. Then, the transmitted and reflected light are related by the following equation:

$$\begin{pmatrix} E_x^t \\ E_y^t \\ 0 \\ 0 \end{pmatrix} = T^{\text{total}} \begin{pmatrix} E_x^i \\ 0 \\ E_x^r \\ E_y^r \end{pmatrix}. \tag{6.9}$$

The total transfer matrix T^{total} is given by the product of the transfer matrices of Eqs. 6.7, 6.8,

$$T^{total} = T^{I}\left(n_f, n_s, \sigma_{xx}^b, 0\right) T^{B} T^{I}\left(n_v, n_f, 0, \sigma_{xy}^t\right)$$

$$= \frac{1}{2}\begin{pmatrix} 1 + \frac{n_v}{n_s} - \frac{2\alpha\sigma_{xx}^b}{n_s} & -\frac{2\alpha\sigma_{xy}^t}{n_s} & 1 - \frac{n_v}{n_s} - \frac{2\alpha\sigma_{xx}^b}{n_s} & -\frac{2\alpha\sigma_{xy}^t}{n_s} \\ \frac{2\alpha\sigma_{xy}^t}{n_s} & 1 + \frac{n_v}{n_s} - \frac{2\alpha\sigma_{xx}^b}{n_s} & \frac{2\alpha\sigma_{xy}^t}{n_s} & 1 - \frac{n_v}{n_s} - \frac{2\alpha\sigma_{xx}^b}{n_s} \\ 1 - \frac{n_v}{n_s} + \frac{2\alpha\sigma_{xx}^b}{n_s} & \frac{2\alpha\sigma_{xy}^t}{n_s} & 1 + \frac{n_v}{n_s} + \frac{2\alpha\sigma_{xx}^b}{n_s} & \frac{2\alpha\sigma_{xy}^t}{n_s} \\ -\frac{2\alpha\sigma_{xy}^t}{n_s} & 1 - \frac{n_v}{n_s} + \frac{2\alpha\sigma_{xx}^b}{n_s} & -\frac{2\alpha\sigma_{xy}^t}{n_s} & 1 + \frac{n_v}{n_s} + \frac{2\alpha\sigma_{xx}^b}{n_s} \end{pmatrix}.$$

$$\tag{6.10}$$

By substituting Eqs. 6.10–6.9,

$$2E_x^t = \left(1 + \frac{n_v}{n_s} - \frac{2\alpha\sigma_{xx}^b}{n_s}\right) E_x^i + \left(1 - \frac{n_v}{n_s} - \frac{2\alpha\sigma_{xx}^b}{n_s}\right) E_x^r - \frac{2\alpha\sigma_{xy}^t}{n_s} E_y^r,$$

$$2E_y^t = \frac{2\alpha\sigma_{xy}^t}{n_s} E_x^i + \frac{2\alpha\sigma_{xy}^t}{n_s} E_x^r + \left(1 - \frac{n_v}{n_s} - \frac{2\alpha\sigma_{xx}^b}{n_s}\right) E_y^r,$$

$$0 = \left(1 - \frac{n_v}{n_s} + \frac{2\alpha\sigma_{xx}^b}{n_s}\right) E_x^i + \left(1 + \frac{n_v}{n_s} + \frac{2\alpha\sigma_{xx}^b}{n_s}\right) E_x^r + \frac{2\alpha\sigma_{xy}^t}{n_s} E_y^r,$$

$$0 = -\frac{2\alpha\sigma_{xy}^t}{n_s} E_x^i - \frac{2\alpha\sigma_{xy}^t}{n_s} E_x^r + \left(1 + \frac{n_v}{n_s} + \frac{2\alpha\sigma_{xx}^b}{n_s}\right) E_y^r.$$

$$\tag{6.11}$$

Finally, we get the Faraday rotation angle as,

$$\tan\theta_F = \frac{E_y^t}{E_x^t} = \frac{2\alpha\sigma_{xy}^t}{n_s + n_v + 2\alpha\sigma_{xx}^b}. \tag{6.12}$$

On the other hand, for the Kerr rotation measurement (Fig. 6.12b), the incident light comes from the substrate side. In this case, the total transfer matrix is rewritten as

$$T^{total} = T^{I}\left(n_f, n_v, 0, \sigma_{xy}^t\right) T^{B} T^{I}\left(n_s, n_f, \sigma_{xx}^b, 0\right)$$

$$= \frac{1}{2}\begin{pmatrix} 1 + \frac{n_s}{n_v} - \frac{2\alpha\sigma_{xx}^b}{n_v} & -\frac{2\alpha\sigma_{xy}^t}{n_v} & 1 - \frac{n_s}{n_v} - \frac{2\alpha\sigma_{xx}^b}{n_v} & -\frac{2\alpha\sigma_{xy}^t}{n_v} \\ \frac{2\alpha\sigma_{xy}^t}{n_v} & 1 + \frac{n_s}{n_v} - \frac{2\alpha\sigma_{xx}^b}{n_v} & \frac{2\alpha\sigma_{xy}^t}{n_v} & 1 - \frac{n_s}{n_v} - \frac{2\alpha\sigma_{xx}^b}{n_v} \\ 1 - \frac{n_s}{n_v} + \frac{2\alpha\sigma_{xx}^b}{n_v} & \frac{2\alpha\sigma_{xy}^t}{n_s} & 1 + \frac{n_s}{n_v} + \frac{2\alpha\sigma_{xx}^b}{n_v} & \frac{2\alpha\sigma_{xy}^t}{n_v} \\ -\frac{2\alpha\sigma_{xy}^t}{n_v} & 1 - \frac{n_s}{n_v} + \frac{2\alpha\sigma_{xx}^b}{n_v} & -\frac{2\alpha\sigma_{xy}^t}{n_v} & 1 + \frac{n_s}{n_v} + \frac{2\alpha\sigma_{xx}^b}{n_v} \end{pmatrix},$$

$$\tag{6.13}$$

in which n_s is just interchanged with n_v for Eq. 6.13. Thus, the Kerr rotation angle is described by

$$\tan \theta_K = \frac{E_y^r}{E_x^r} = \frac{4\alpha n_s \sigma_{xy}^t}{n_s{}^2 - n_v{}^2 + 4\alpha n_v \sigma_{xx}^b + 4\alpha^2 \left(\sigma_{xx}^b{}^2 + \sigma_{xy}^t{}^2\right)}. \tag{6.14}$$

To better clarify the contribution of the bottom surface transport (σ_{xx}^b), we plot θ_F and θ_K versus σ_{xx}^b in Fig. 6.13 by substituting $n_v = 1$ and $n_s = 3.46$ (for InP substrates) and $\sigma_{xy}^t = 1/2(e^2/h)$ to Eqs. 6.12, 6.13. Even in the presence of finite σ_{xx}^b, θ_F and θ_K take almost constant values. For the typical semi-magnetic TI films with $\sigma_{xx}^b \sim 1(e^2/h)$, the error from the half-quantized value with $\sigma_{xx}^b \sim 0$ amounts to only 0.3% (0.3%) for the Faraday (Kerr) rotation. Even for the case of $\sigma_{xx}^b \sim 4(e^2/h)$, which is the largest value presented in Fig. 6.7, the error is only 1.3% (1.1%). Thus, the Faraday (Kerr) rotation is insensitive to σ_{xx}^b in the semi-magnetic TI. Therefore, the Faraday (Eq. 6.12) and Kerr rotation (Eq. 6.14) angles can be reasonably approximated to,

$$\theta_F \approx \tan^{-1}\left(\frac{\alpha}{n_s + 1}\right), \tag{6.15}$$

$$\theta_K \approx \tan^{-1}\left(\frac{2\alpha n_s}{n_s{}^2 - 1}\right). \tag{6.16}$$

Compared with the quantized Faraday and Kerr rotation angles in the Q(A)H states as calculated previously [3, 4, 18],

$$\theta_F^Q = \tan^{-1}\left(\frac{2\alpha}{n_s + 1}\right), \tag{6.17}$$

$$\theta_K^Q = \tan^{-1}\left(\frac{4\alpha n_s}{n_s{}^2 - 1 + 4\alpha^2}\right) \approx \tan^{-1}\left(\frac{4\alpha n_s}{n_s{}^2 - 1}\right), \tag{6.18}$$

the calculated rotation angles for the semi-magnetic TI films give rise to the half of the integer-quantized rotations θ_F^Q and θ_K^Q, corresponding to our experimental results.

6.5.2　Calculation of Hall Conductivity in Semi-magnetic Topological Insulator

We show that the half quantized Hall conductivity is not maintained in the presence of the electronic hybridization between the top and bottom surface states.

6.5.2.1　Calculation

We use an effective continuum model for the TI surface states (in x-y plane) including the top and bottom surface degrees of freedom [19]. The Hamiltonian for the

Fig. 6.13 Calculated σ_{xx}^b dependence on θ_F and θ_K using Eqs. 6.12 and 6.14. The broken lines indicate the θ_F and θ_K for $\sigma_{xx}^b = 0$. The insets are magnified views for θ_F and θ_K

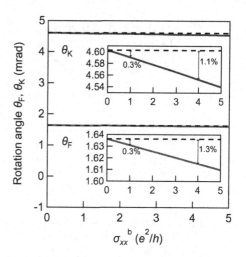

wavevector $k = (k_x, k_y)$, which is expressed in the $| t \uparrow \rangle$, $| t \downarrow \rangle$, $| b \uparrow \rangle$ and $| b \downarrow \rangle$ bases (t and b denote the top and bottom surface states, $\uparrow$ and $\downarrow$ represent the spin up and down states, respectively), can be written as

$$H(k_x, k_y) = \begin{pmatrix} \Delta_t & k_y - ik_x & m & 0 \\ k_y + ik_x & -\Delta_t & 0 & m \\ m & 0 & \Delta_b & -k_y + ik_x \\ 0 & m & -k_y - ik_x & -\Delta_b \end{pmatrix}, \qquad (6.19)$$

where m is a parameter for the hybridization gap, and Δ_t and Δ_b are magnetically-induced gaps for the top and bottom surfaces, respectively. Here, we set $\Delta_b = 0$ to express the situation of the semi-magnetic TI structure. By solving the eigen-equation for this Hamiltonian, the eigen-energies can be analytically obtained,

$$E = \pm \sqrt{\frac{\Delta_t^2}{2} + m^2 + k^2 \pm \sqrt{\frac{\Delta_t^2}{4} + m^2}}, \qquad (6.20)$$

where $k = \sqrt{k_x^2 + k_y^2}$. From the eigen-functions $|u_{nk}\rangle$, corresponding to the eigen-energies E_n, we compute the Berry connection and curvature defined as $a_n(k) = i \langle u_{nk} | \nabla_k | u_{nk} \rangle$ and $b_n(k) = \nabla_k \times a_n(k)$, respectively. Then, the Berry curvature contribution of the Hall conductivity σ_{xy} arising from the above 4 bands is given by [20],

$$\sigma_{xy} = \frac{1}{2\pi} \sum_{n=1}^{4} \iint_{-\infty}^{\infty} d^2k \, f_n(k) \, b_n(k), \qquad (6.21)$$

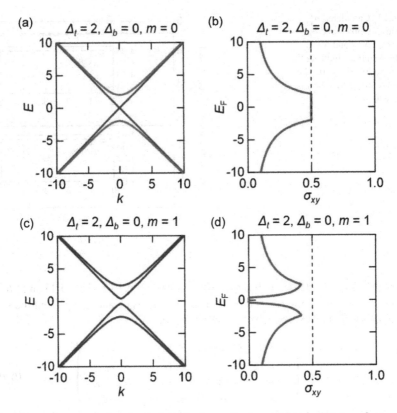

Fig. 6.14 a, c Low-energy bands calculated for the continuum model including $\Delta_t = 2, m = 0$ (**a**) and $\Delta_t = 2, m = 1$ (**c**). **b, d** Calculated σ_{xy} versus E_F corresponding to the band structures in (**a**) and (**c**), respectively

where $f_n(k) \equiv 1/(e^{(E_n - E_F)/k_B T} + 1) \to \theta(E_F - E_n)$ (when $T \to 0$) is the Fermi distribution function and θ is the Heaviside step function.

In Fig. 6.14a, we show the low-energy bands at the zero-temperature limit, where the hybridization is not included ($m = 0$). The red lines indicate the top surface state which has the magnetic gap ($\Delta_t = 2$), whereas the blue lines indicate the bottom surface state which has no gap ($\Delta_b = 0$). Then, the calculated σ_{xy} is shown in Fig. 6.14b, exhibiting a plateau structure with the half-integer quantized value when E_F locates within the magnetic gap.

We then include the hybridization term ($m \neq 0$). Figure 6.14c shows the numerically calculated bands including finite $m(= 1 = \Delta_t/2)$ that exhibits a gap near $E = 0$. This gap is a trivial one because σ_{xy} becomes zero when E_F is within the gap as shown in Fig. 6.14d, indicating that the system no longer exhibits the half-integer quantization in the presence of hybridization. We note that, when E_F intersects the bands, σ_{xy} has a non-zero value (but lower than the half-quantized value). Also, there are peaks at the band edges of the highest and lowest bands. These features are indeed observed experimentally as discussed below.

6.5.2.2 Experiment

We have examined 7-nm-thick films (Fig. 6.15a), which possess sizeable hybridization gaps comparable to their magnetization gaps [21]. Figure 6.15b shows the V_g dependence of σ_{xy} in the film of $x = y = 0.88$ which we have presented in our previous publication [6], where E_F is supposed to be above the magnetic gap of the top surface state and is slightly below the Dirac point for the bottom surface state. In contrast to the half-integer quantization in 10-nm-thick films presented in the main text, the half-quantized σ_{xy} does not appear at $\mu_0 H = 0$ T while the $\nu = 0$ and 1 QH plateaus emerge at $\mu_0 H = 14$ T.

Furthermore, we observe a dip at around $V_g = 0$ V. Within this dip, σ_{xx} also takes a minimum value (Fig. 6.15c). This indicates the hybridization-induced trivial gap as expected from the numerical results shown in Fig. S3d. We have additionally checked another sample in which we finely tune E_F within the magnetic gap of the top surface ($x = 0.73$, $y = 0.85$), leading to an even more insulating state (Figs. 6.15d, e). We

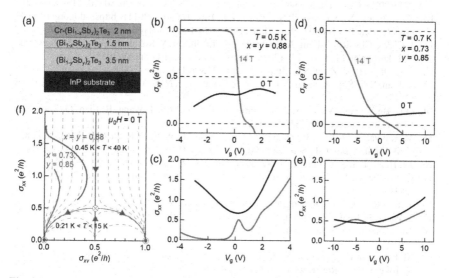

Fig. 6.15 **a** Schematic layout of 7-nm-thick semi-magnetic TI films studied in (**b**–**f**). V_g dependence of σ_{xy} and σ_{xx} in a semi-magnetic TI film ($x = y = 0.88$) at $T = 0.5$ K and $\mu_0 H = 0$, 14 T, reproduced from [6]. **d, e** V_g dependence of σ_{xy} and σ_{xx} in a semi-magnetic TI film ($x = 0.73$, $y = 0.85$) at $T = 0.7$ K and $\mu_0 H = 0$, 9 T. **f** T-driven flows of (σ_{xy}, σ_{xx}) at $\mu_0 H = 0$ T. The broken lines are reproduced from [9]

then map $(\sigma_{xy}(T), \sigma_{xy}(T))$ at the most insulating V_g (Fig. 6.15f) for the two samples, flowing to the trivial insulating fixed point (0, 0). In conclusion, thickening the films so as to suppress the surface hybridization is required to observe the half quantized σ_{xy}.

References

1. T. Morimoto, Y. Hatsugai, H. Aoki, Phys. Rev. Lett. **103**(11), 116803 (2009)
2. W.K. Tse, A.H. MacDonald, Phys. Rev. Lett. **105**(5), 057401 (2010)
3. J. Maciejko, X.L. Qi, H.D. Drew, S.C. Zhang, Phys. Rev. Lett. **105**(16), 166803 (2010)
4. K.N. Okada, Y. Takahashi, M. Mogi, R. Yoshimi, A. Tsukazaki, K.S. Takahashi, N. Ogawa, M. Kawasaki, Y. Tokura, Nat. Commun. **7**, 12245 (2016)
5. X.L. Qi, T.L. Hughes, S.C. Zhang, Phys. Rev. B **78**(19), 195424 (2008)
6. R. Yoshimi, K. Yasuda, A. Tsukazaki, K. Takahashi, N. Nagaosa, M. Kawasaki, Y. Tokura, Nat. Commun. **6**, 8530 (2015)
7. R. Shimano, G. Yumoto, J. Yoo, R. Matsunaga, S. Tanabe, H. Hibino, T. Morimoto, H. Aoki, Nat. Commun. **4**, 1841 (2013)
8. A.M.M. Pruisken, Phys. Rev. Lett. **61**, 1297 (1988). https://doi.org/10.1103/PhysRevLett.61.1297. https://link.aps.org/doi/10.1103/PhysRevLett.61.1297
9. B.P. Dolan, Nucl. Phys. B **554**(3), 487 (1999). https://doi.org/10.1016/S0550-3213(99)00326-0. http://www.sciencedirect.com/science/article/pii/S0550321399003260
10. K. Nomura, N. Nagaosa, Phys. Rev. Lett. **106**, 166802 (2011). https://doi.org/10.1103/PhysRevLett.106.166802. https://link.aps.org/doi/10.1103/PhysRevLett.106.166802
11. J. Checkelsky, R. Yoshimi, A. Tsukazaki, K. Takahashi, Y. Kozuka, J. Falson, M. Kawasaki, Y. Tokura, Nat. Phys. **10**, 731 (2014). https://doi.org/10.1038/nphys3053
12. X.L. Qi, R. Li, J. Zang, S.C. Zhang, Science **323**(5918), 1184 (2009)
13. B. Lian, X.Q. Sun, A. Vaezi, X.L. Qi, S.C. Zhang, Proc. Natl. Acad. Sci. USA **115**(43), 10938 (2018)
14. Q.L. He, L. Pan, A.L. Stern, E.C. Burks, X. Che, G. Yin, J. Wang, B. Lian, Q. Zhou, E.S. Choi et al., Science **357**(6348), 294 (2017)
15. M. Kayyalha, D. Xiao, R. Zhang, J. Shin, J. Jiang, F. Wang, Y.F. Zhao, R. Xiao, L. Zhang, K.M. Fijalkowski et al., Science **367**(6473), 64 (2020)
16. J.J. He, T. Liang, Y. Tanaka, N. Nagaosa, Commun. Phys. **2**(1), 1 (2019)
17. G. Széchenyi, M. Vigh, A. Kormányos, J. Cserti, J. Phys. Condens. Matter **28**(37), 375802 (2016)
18. W.K. Tse, A.H. MacDonald, Phys. Rev. B **84**(20), 205327 (2011)
19. J. Wang, B. Lian, S.C. Zhang, Phys. Scr. **2015**(T164), 014003 (2015)
20. N. Nagaosa, J. Sinova, S. Onoda, A.H. MacDonald, N.P. Ong, Rev. Mod. Phys. **82**(2), 1539 (2010)
21. M. Kawamura, M. Mogi, R. Yoshimi, A. Tsukazaki, Y. Kozuka, K.S. Takahashi, M. Kawasaki, Y. Tokura, Phys. Rev. B **98**(14), 140404 (2018)

Chapter 7
Summary

In this thesis, we have explored quantized phenomena in magnetic topological insulators (TIs). Our heterostructure engineering strategies based on the magnetic modulation doping and the magnetic proximity coupling have opened a way to observe and stabilize the quantum transport and magneto-optical phenomena. Furthermore, in contrast to the standard uniformly doped system, we magnetically control the top/bottom surface degrees of freedom, which dramatically enriches the variation of quantized phenomena in magnetic TIs.

In Chap. 3, we have developed a Cr magnetic modulation doping technique for $(Bi, Sb)_2Te_3$ to enhance the exchange interaction between the surface states and the magnetic ions as well as to reduce disorder in bulk states. Comparison among the uniformly doped and modulation-doped samples through the gate voltage and temperature dependences, we find the enhancement of the magnetization-induced gap for the surface states. By optimizing the density of Cr ions, we observe the QAH effect up to 2 K, which is orders of magnitude higher than the uniformly doped TIs.

In Chap. 4, we have grown Te-based van der Waals ferromagnetic insulators $Cr_2Ge_2Te_6$ and $Cr_2Si_2Te_6$ thin films, which enables us to grow sandwich structures with a $(Bi, Sb)_2Te_3$ TI layer. Through the efficient interfacial exchange interaction between the topological surface state and the adjacent ferromagnetic insulator layers, we observed the signature of the QAH state driven by the magnetic proximity coupling. Furthermore, by using such a topological interface state with a spin-momentum locking property, we have realized the electrical current induced magnetization switching. These results indicate the possibility of electrical switching of the QAH effect and of the emergence of exotic interfacial phenomena with magnetism by utilizing a wide variety of magnetic insulator layers.

In Chap. 5, we have investigated topological phase transitions from the QAH state. By tuning the magnetization gap via the rotation of an external magnetic field, namely the rotation of magnetization direction, we observe the transition from the QAH

M. Mogi, *Quantized Phenomena of Transport and Magneto-Optics in Magnetic Topological Insulator Heterostructures*, Springer Theses,
https://doi.org/10.1007/978-981-19-2137-7_7

107

insulator to the trivial insulator owing to the competing surface hybridization gap in a range of film thickness (7 to 8 nm). Furthermore, in the modulation-doped sandwich heterostructures endowed with asymmetries, such as vertical asymmetric Cr doping and V/Cr doping for the top/bottom surfaces, a different type of the insulator phases, termed axion insulator, appears with the annihilation of chiral edge states. This axion insulator phase may host a quantized magnetoelectricity, which is expected to be observed in the near future. These results establish the independent control of top and bottom surface Dirac electrons with broken time-reversal symmetry.

In Chap. 6, we have explored exotic low-energy electrodynamics in a 3D TI. In the QAH state, we observed the quantized Faraday and Kerr rotations in THz frequency. Hence, we have established that the terahertz magneto-optical spectroscopy is a powerful technique to probe the dynamical topological responses in a low-frequency region. Furthermore, we use this technique to the modulation-doped semi-magnetic TI, where one surface is magnetically gapped while the other surface remains gapless. We observed half-quantized Faraday and Kerr rotations at zero magnetic field. This half-integer quantization is also observed in dc-transport Hall conductivity, which supports the bulk-surface correspondence. This result confirms the parity anomaly of 2D Dirac fermions on the surface of a TI in terms of quantum field theory.

Thus, our heterostructure engineering based on magnetic TIs has realized the high-quality magnetic topological surface states and enabled us to control spatial degrees of freedom of the surface states hosting 2D spin-polarized Dirac electrons, leading to various quantized phenomena. Toward even higher temperature realization of the quantized states, control of disorders in topological insulators is a key issue. In Cr-doped $(Bi, Sb)_2 Te_3$, the Bi/Sb carrier compensation for Te deficiency/anti-site defects and random magnetic impurities lead to the significant spatial inhomogeneities for the Dirac point and the magnetization. Recently, the introduction of insulating buffer layers with low donor doping has achieved low-carrier density and high-mobility TIs owing to the reduction of defects [1, 2]. As well as the magnetic proximity effect from intrinsic ferromagnetic insulators as studied in Chap. 4, synthesis of intrinsic magnetic TIs is a promising route to eliminate the randomness in magnetic moments [3–7]. Furthermore, our heterostructure studies would be helpful for further exploration of heterostructure designing, such as the introduction of a superconducting proximity effect and choices of magnetic insulators including antiferromagnets and spin non-collinear magnets, for future electrical/optical applications based on topological surface states.

References

1. N. Koirala, M. Brahlek, M. Salehi, L. Wu, J. Dai, J. Waugh, T. Nummy, M.G. Han, J. Moon, Y. Zhu, D. Dessau, W. Wu, N.P. Armitage, S. Oh, Nano Lett. **15**(12), 8245 (2015)
2. M. Salehi, H. Shapourian, I.T. Rosen, M.G. Han, J. Moon, P. Shibayev, D. Jain, D. Goldhaber-Gordon, S. Oh, Adv. Mater. **31**, 1901091 (2019)
3. J. Li, Y. Li, S. Du, Z. Wang, B.L. Gu, S.C. Zhang, K. He, W. Duan, Y. Xu, Sci. Adv. **5**(6), eaaw5685 (2019)

4. M. Otrokov, I. Klimovskikh, H. Bentmann, D. Estyunin, A. Zeugner, Z. Aliev, S. Gaß, A. Wolter, A. Koroleva, A. Shikin et al., Nature **576**(7787), 416 (2019)
5. Y. Gong, J. Guo, J. Li, K. Zhu, M. Liao, X. Liu, Q. Zhang, L. Gu, L. Tang, X. Feng et al., Chin. Phys. Lett. **36**(7), 076801 (2019)
6. Y. Deng, Y. Yu, M.Z. Shi, J. Wang, X.H. Chen, Y. Zhang (2019). arXiv:1904.11468
7. C. Liu, Y. Wang, H. Li, Y. Wu, Y. Li, J. Li, K. He, Y. Xu, J. Zhang, Y. Wang (2019). arXiv:1905.00715

Printed in the United States
by Baker & Taylor Publisher Services